WHALES, DOLPHINS AND PORPOISES

Ronald M. Lockley

WHALES, DOLPHINS AND PORPOISES

with illustrations by
Elizabeth Sutton

DAVID & CHARLES
Newton Abbot   London

British Library Cataloguing in Publication Data

Lockley, Ronald Mathias
Whales, dolphins and porpoises.
1. Cetacea
I. Title
599'.5 QL737.C4

ISBN 0-7153-7731-0

Set in 11 on 12pt Plantin
by ABM Typographics Limited, Hull
and printed in Great Britain
by Alden Press, Oxford
for David & Charles (Publishers) Limited
Brunel House Newton Abbot Devon

Contents

Preface

The dolphin carrieth a loving affection not only unto man, but also to music; delighted is he with harmony in song, but especially with the sound of the water instrument, or such kind of pipes. Of man he is nothing afraid, neither avoideth him as a stranger; but of himself meeteth their ships, plaieth and disporteth himself and fetcheth a thousand frisks and gambols before them.

Pliny the Younger, AD 61-113

Dolphins are small whales, and porpoises—a popular name in the United States—are small dolphins. All are whales, belonging to the same aquatic mammal order of *Cetacea* (from the Latinised Greek *kêtos*, whale). They are the cetaceans, a sonorous sea-sounding term which is used collectively here, a name which to me partakes of the hiss and surf and song of the ocean which I can hear and see from my window. For I count myself fortunate in a view which opens 'on the foam of perilous seas, in faery lands forlorn': to be exact, the great Hauraki Gulf of New Zealand, whale- and dolphin-haunted and large enough to provide anchorage for all the navies of the world.

Ringed by a chain of large and small islands, the Gulf lies in the migratory track of the giant humpback, pilot, killer and false killer whales and other cetaceans. As I write it is autumn here, and many are moving north, summer-fat and full-fed on the boundless seafood of the cool Antarctic and sub-Antarctic oceans to the south.

Often those to come closest inshore are the ungainly-looking humpbacks, with their unsmooth, warty appearance and the longest flippers of any cetacean. From the open Pacific coasts and heavy surf of the Bay of Plenty, they pass into the Gulf through the rippling tides of the Colville Channel, finding a calm mirror under the high land of the Firth of Thames and Great Barrier Island.

In small family units of a pair, a pair with one calf, or a pair with a calf and the adolescent born over two years earlier, they seem to enjoy the smooth iridescent water. They move their prodigious bulk slowly, at an easy two or three knots, and call and sing to each other.

They are much scarcer today. I sometimes wonder—do they come so close inshore because they know they now enjoy total protection from the human enemies who once killed hundreds of them annually in these sheltered waters ? The last time they were slaughtered and their oil tried out at Whangaparapara Bay, on the inside coast of Great Barrier Island,

An abandoned whaling station on Great Barrier Island, New Zealand
(Auckland Star)

was in 1963. Or is it that they have forgotten that murder and are too trusting by nature?

All the great whales have been approachable and peaceable on their first encounter at sea with man; it is only from the agonising experience of being hunted, and wounded with harpoon, lance or bullet that they become wary and flee at the appearance of motor boats and the sound of engines. As many filming skindivers have lately shown, you can swim close to almost any of the great cetaceans, even the notorious killer whale, and touch (occasionally ride upon the back of) an individual, provided you move gently and silently. The cetacean will look you in the eye if the water is clear; in any case even in murky water it will know much about you by measuring you accurately with its sonar scanner, which—it is now believed—informs it of your size, shape, density, heartbeat rate and attitude to other living creatures. But more about this in-depth scanning device in the chapter on sonar.

Having decided you are harmless, some cetaceans, especially the young and playful adolescents, will show curiosity and follow the skin-diver. With very little encouragement some species will enjoy a game with you under water. Pliny and Aristotle were right—these lively sea-mammals carry a loving affection unto man.

Sailing silently in *Siesta*, my twenty-three-foot sloop, off Great Barrier Island, I enjoy my first encounter with a humpback family at short range. Suddenly the nearest leaps quite clear of the surface, an amazing feat for an animal weighing about 40 tons which only a moment ago had been quietly swimming at a few knots. He slaps back into the sea, a belly or side flop—out of sheer *joie de vivre?* Almost at once the other adult (probably the cow—she is larger, perhaps 15m (50ft) long) shoots into the air, tossing off a mane of white water from her broad lumpy forehead, causing an enormous wave as she falls backwards towards where I had last seen her calf. The ripples of the adults' splashes have not travelled far before the calf makes a skywards leap in a very fair imitation of its parents. It soars at least half a metre clear of the bubbling white surface, a smoother, darker object than the old ones, and only one-third their size.

There is not enough breeze for *Siesta*, who lives up to her name—she is slow in a light zephyr, a roomy twin-keeler which can lie with ease at low tide on the sand of many a hidden uninhabited cove during our timeless exploration of this sunlit gulf. The humpbacks, despite their apparently leisurely playful progress, are leaving us behind. They are swimming north between Great and Little Barrier Islands. I start the auxiliary engine, anxious to catch up with their spectacular leaping forms, their long flippers, silvery beneath, flashing in the sun as they somersault backwards, plunging with open arms upon the sea, to disappear with a tail-slap which echoes back a few seconds later from tall island cliffs. Rising once more, revolving that great back at the surface, then like other whales after a good blow, the humpback lobs its broad tail-flukes into the air as it makes a steep dive.

The humpbacks are making their annual three-month journey towards the equator, a happy unit as they swim, often so close together that their bodies frequently touch. The calf glides above its mother, its flippers caressing her humpy back; usually they breach and breathe in unison, the calf slipping to one side to do so every twenty seconds or so, then resuming its position above mother's back as they submerge. Father (or is it auntie—a pregnant cow?) swims close by and sometimes the calf is alongside, guarded between both adults.

Would that we could follow the devoted family away from the approaching southern winter to warm trade winds and equatorial waters where they will enjoy more high-spirited play and tuneful songs, where the adults indulge in their face-to-face mating ritual, so seldom observed but now accurately recorded. And new calves are born in a warmer sea.

From much evidence of family togetherness, it was thought that humpback, blue, fin and most of the huge baleen whales, as well as many smaller toothed cetaceans, were monogamous, that some may pair for life, which can last as long as a human life. Whale-men well know and take advantage of this devotion: when one of a pair or family is wounded, the other(s) will not abandon the stricken animal—even perhaps after death. The hunter knew that having harpooned one, the entire family was vulnerable as the others swam to the aid of their comrade or child.

Even with the largest of the toothed whales, that deepest diver of all cetaceans, the sperm whale or cachalot, which is not monogamous, but adopts a harem system of many wives (faithful to a master bull during the seasons of his ascendency), there is instant closing of ranks and a rush to help any member of the school calling in distress, regardless of sex, age or other social position in the hierarchy.

Near where I live this was dramatically, tragically demonstrated on two terrible occasions. On the stormy night of 29 October 1974, seventy-five adult and near-adult sperm whales, including many pregnant females (but no small calves) stranded themselves along a mile of lonely west-coast beach under the dunes of Muriwai Sands, far from sight of any human habitation. They were not noticed until low tide early next morning. Not one of these immense animals survived to breathe by the time the evening tide returned to lap their blistered bodies. They had died one by one in sight of the surf of the Tasman Sea which beats ceaselessly upon this shore during the prevailing westerly winds. It was quite impossible to move any one of these giants, the smallest of which would weigh five, and the largest male up to sixty tons. I counted at least forty females, many had been pregnant enough to abort their calf after death, under the pressure of the gases of decomposition which generate swiftly in the huge warm body.

They had died not from lack of air, but from lack of cooling water. Their vast weight, unsupported by water of equal pressure when the tide ebbed, had no doubt pressed uncomfortably, unequally, upon the blood–engorged organs of heart, lungs, liver and kidneys. Possibly in an endeavour to relieve this pressure all were lying on one side, one flipper in the air. But without water to cool the blood circulating through the skin and limbs these unfortunate whales in fact had suffocated from the increasing internal temperature generated by the uncooled blood. Whales are quite unable to reduce body-heat by perspiring or panting. Whale-hunters will tell you that the sooner you process a large whale after killing it, the better will be the profit; left unopened for half a day out of the water, the great carcase begins to 'cook' in its internal heat, and the quality of the oil, blubber and other products rapidly deteriorates through the chemical changes of decomposition.

On 1 April 1978, a school of false killer whales swam in their usual close formation through the narrow entrance of the shallow, sheltered, vast Manukau Harbour, which fringes the southern suburbs of Auckland, New Zealand's largest city. These small all-black whales, between 3.6m (12ft) and 5.4m (18ft) long, and weighing up to 2,000kg (4,400lb), but usually much less, likewise chose the night to strand on the falling tide. Strange cries had been heard by a night fisherman, who saw black forms in the distance, and thought they were part of a herd of cattle—not an uncommon sight wading in the edge of the sea and adjacent fields.

Those cries were the distress calls of the first whales stranded. They did not die at once; their narrow bodies, resting easily on the muddy

A male sperm whale stranded on Muriwai Sands near Auckland, New Zealand, in October 1974, one of 75 which died on this occasion. The lower jaw has been sawn off by souvenir-hunters—the teeth fetch a good price as long-lasting ivory. The gases of decomposition have forced the extrusion of the penis; in a pregnant female, the gases are liable to cause the abortion of the foetus after death (Ronald M. Lockley)

foreshore, did not overheat during the cool night hours. Unfortunately they had chosen the little-visited southern part of the Manukau Harbour which is sparsely inhabited, and their predicament was not realised in time for help to be organised until late the next day. Those stranded high upon this enormous beach meanwhile were scorched by the strong sunlight, their thin black outer skin peeling and curling in strips—like carbon-paper under heat. About fifty died before the returning tide reached them.

A small band of whale-lovers worked all night to push back into the sea while the tide was in as many as possible of some two hundred of the school which pressed inshore, seeking to come to the help of those stranded and still alive and making faint bleeping calls.

For a while we thought we had succeeded in persuading some of these beautiful sleek black creatures to swim away. They were amazingly passive and gentle when we walked each one with our arms around 'shoulders' and 'thighs' through the waist-high water and held them upright until their sense of balance was restored. They seemed to welcome our company and assistance; and once two or three were gathered together in a bunch they moved off, side by side, fins touching, into the dusky distance, snorting periodically.

But as the tide rapidly ebbed and dawn came we could see dozens had returned and were stranded again. After their swim in the sea had refreshed them they still would not abandon those crying in distress ashore. Again it was a warm late summer day, with no wind. The sun burned down, scorching and suffocating those whales still alive, and the sea ebbed too far for helpers to carry water to cool the stranded ones.

The combination of circumstances on this occasion was against saving most of the whales by human action. Such a huge stranding had never occurred before in New Zealand; the authorities were totally unprepared, with no knowledge of what to do, nor of who was responsible—was it the Manukau Harbour Master, the Fisheries Department, or the Marine Division (of the Ministry of Transport)? The farmer through whose fields lay the shortest access from a public road became angry with the numbers of trespassing cars loaded with officials, sightseers and genuine helpers, and locked his gates, thereby unfortunately denying the whales further much-needed assistance to prevent the rest of the school, ranging close inshore, from stranding on the falling tide. Too late a plan was put into action, recommended by officials of Project Jonah (that world-wide and admirable whale-protection society), to kill humanely all the stranded whales, and thus silence their distress calls. From long experience of strandings this is believed to be the only way to persuade the rest of the school to swim away, for they will rarely do so as long as they can hear even one stranded companion calling in distress.

Police permission had first to be obtained to use rifles; this caused further delay. But many whales were already dead, and all through the ebbing tide that second day others were stranding here and there over three miles of the vast estuary; some which did so for the second, perhaps third time, were weak and ready to die.

At low tide that afternoon I counted 253; all were false killer whales *Pseudorca crassidens*. Seventy-seven nearest to the high-water mark had been shot by a Fisheries Officer, and Animal Health Officers had opened their carotid arteries to ensure that none remained alive. Of forty bodies examined by Project Jonah members, twenty were females of an average length of 4m (13ft); a large cow which was dissected on the beach was carrying a calf 1.5m (5ft) long. The twenty males examined averaged 4.7m (15½ft).

A terrible sight, and a great sadness filled me as I walked for four hours along the low-water mark, trying to help the last to be stranded; by this time a helicopter had begun to lift away the carcasses of the first to die, at the top of the beach—to be processed at a meat factory inland.

The loneliest, last and lowest on the strand was a large bull lying awash half a mile from the nearest corpse. He startled me with an explosive exhalation through the blow-hole at the top of his head. For I had been about to measure him, thinking he must be dead; his body

False killer whale (Jen & Des Bartlett/Bruce Coleman)

was terribly lacerated, the skin peeling in the dry blistering sunlight to reveal areas of blubber veined with scarlet blood-vessels.

When I gently waded in to touch his face, one eye opened wide for a staring second. The uppermost flipper (he was lying on the other) waved a pitiful greeting.

'My poor fellow—you are still alive ?'

Only he could hear my silly exclamation; the prostrate whale closed that eye, reproachfully I thought, as if he had said, 'Idiot, of course I'm alive! Splash me some more sea. It's comforting. Can't you see my skin is burning dry ?'

'I fear the tide is still ebbing,' I said aloud, for there was not another human in view, only a sight-seeing plane passing high overhead. 'But what can I do ? You must weigh at least two tons for your 5.5 metre length. I can't possibly shove you back into the sea . . .'

But he made no attempt to help me when I lifted his tail momentarily, opening that eye briefly again, and waving a flipper negatively. He was too firmly grounded, too heavy to budge.

As long as water remained accessible I bathed his burned body with capfuls. This seemed—I liked to believe—to ease him. He snorted at regular intervals—to be precise every 20 seconds by my watch. He seemed to enjoy my gentle stroking of his face, which was meant to reassure him that I was giving out all my heart and sympathy to him, while vainly praying that somehow he could be rescued back to the freedom of the ocean.

'At least I can stay with you as long as possible,' I promised him when next he gave that prodigious snorting sigh, and opened that sensitive eye to stare at me, flicking the lids a little, then puckering them shut—like a tired old grandfather longing to sleep.

'We all have to die', he seemed to be reminding me, flipper twitching as if to emphasise the point. 'You, too . . .'

'But why this mass suicide ? I don't understand—we know so little about your way of life—almost nothing. Is it really because one of your tribe had got into difficulties, or was sick, and you had gone to the rescue ? And was his or your echo-location scanner inadequate on this smooth shore where you could hear no waves breaking on the soft mud to warn you of danger ?'

'We whales—our species at least—stick together as we roam the seas of all the world,' he seemed to be telling me, between those loud sighs, beginning to be longer apart in time. 'Our family parties build up to hundreds. But if only one gets into trouble we mass together to defend a member of our tribe—you call it school.

'We lift our sick to the surface to keep them breathing. We have midwives to provide the same service to the new-born . . . Phoo-oosh! I am very tired . . .'

'But why come into this shallow harbour at all ? It's already polluted, away to the north-east, by the main Auckland sewage outfall.'

There were eight dolphins close together and for a moment they disappeared. With breathless expectation we waited and watched for them to surface and perform their feats of pure magic. But nothing could have prepared us for what happened next.

In pairs they emerged from the water, leaving glimmering a silver wake behind them, and soared up and over some invisible hurdle. Their small dark eyes glinting and their soft cream underbellies flashing and reflecting the sun like hazy mirrors. As they leapt every one of our hearts leapt with them. And the flowing evening air was like a heady perfume emitting a different scent for each person, but just as intoxicating to everyone.

Perhaps I should point out that although there were adults here and there along the darkening beach, the dolphins knew that only children can feel the same joy as theirs, only children could feel the same exhilaration—just being alive feeling the air tingling and touching things with the sense of having touched for the first time. That is why as we watched the dolphins the children's spirits soared with indescribable rapture.

The sun slowly disappeared, hovering on the edge of the horizon, bidding farewell. The dolphins leapt once more, one after the other, and waved their tails in farewell, leaving us on the beach, reliving every moment over and over again.

For me the magical feeling I felt when the dolphins came to us will go on forever. I just wish that the whole world could feel what I felt because I know, if it did, nothing would ever be the same again.

Folklore and fact

So is this great and wide sea, wherein are things creeping innumerable, both small and great beasts. There go the ships: there is that leviathan, whom Thou hast made to play therein.

Psalms 104, 25-26

Could the historians and prophets who wrote the Old Testament have known that leviathan, the great whale, 'played' in the sea? Perhaps the information was derived from the pre-Christian writers of classical Greece and the Roman Empire.

The Mediterranean authors and poets of that period wrote lyrically of the whales and dolphins, already known collectively as cetaceans, regarding them as sacred creatures and endowed with a divine intelligence. To many they were reincarnations of the human soul, therefore to kill one unjustly would invite the wrath of the gods. To a few, more factual and prosaic, they were objects of great curiosity and scientific investigation.

The Greek philosopher Aristotle (384–322 BC) was probably the first to record that whales and dolphins are mammals, in his *History of Animals:*

> The dolphin, the whale and all the cetaceans—that is, all that are provided with a blow-hole instead of gills—are viviparous . . . All animals that are . . . viviparous have breasts, as for instance all animals that have hair, such as man and the horse and the cetaceans.

Of the dolphin he wrote much:

> Its period of gestation is ten months [true today for the common dolphin] . . . it brings forth in summer, and never at any other season. The young accompany it for a considerable period, and in fact the creature is remarkable for the strength of its parental affection . . . Some have lived for more than twenty-five and some for thirty years; fishermen nick their tails sometimes and set them adrift again, and by this expedient their ages are ascertained.

Possibly the first record of a study of longevity by marking in a wild animal? Aristotle was a cautious scientist, but, in the fashion of contemporary writers, liked to leaven his factual descriptions by referring to the tales told by others, but without affirming or denying their truth. The soldier-traveller Pliny the Elder (Gaius Plinius Secundus, AD 23–79) on the other hand, collected and embellished these and many other stories at second- and third-hand, creating inaccuracies in his transcriptions—some obviously taken from Aristotle's *Historia animalium*—for insertion in his *Naturalis historia;* stories which have been much quoted and passed down the centuries by many later Latin and early medieval writers.

Pliny, who was killed in the eruption of Vesuvius in AD 79, was a prolific writer, but we had better stick to Aristotle, and note that the stories he relates (at second-hand) of the intelligence of dolphins and their friendliness towards man were for centuries after disbelieved, even scorned, almost down to the present time. I quote two of these which we now know must have been based on fact:

> Many stories are told about the dolphin, indicative of his kind and gentle nature, and manifestations of passionate attachment to boys, in and about Taras, Karia and other places. One story goes that, after a dolphin had been caught and wounded off the coast of Karia, a school of dolphins came into the harbour and stopped there until the fisherman let his captive free; whereupon the school departed.
>
> On one occasion a school of dolphins, large and small, was seen, and two dolphins appeared swimming underneath a little dead dolphin each time it was sinking, and supported it on their backs, trying out of compassion to prevent its being devoured by some predacious fish.

This fits in perfectly with recent observations of dolphins in captivity, which will nudge and keep a dead or dying calf at the surface for hours, even days. Without doubt this is a partly instinctive action, as well as

The Arethusa coin from Syracuse, a decadrachma struck to celebrate the victory at Himera in 480BC. The nymph Arethusa is always accompanied by four dolphins

An early Greek silver coin depicting Taras after being rescued from drowning by dolphins

A chart decoration depicting Poseidon and Amphitrite with dolphins in attendance (Tony Soper)

typical of the protective care of the parturient mother, who is usually aided by another dolphin, ensuring that the new-born calf can start breathing air. From this natal experience the calf learns its survival value. That the parent will go on supporting a dead baby in the hope it will revive is not surprising in this highly sentient sea-mammal, and may be compared with the behaviour and emotional state of the human mother clinging to her child and unwilling to believe that it has died.

Because of this almost human intelligence and behaviour and kindness to man, the dolphins were a favourite motif with the early artists of the golden age of Mediterranean civilisation. They are freely portrayed in sculpture, drawing, painting and on coins. One of the earliest, as well as the most beautiful and accurate, is the decorative mural in colour of a group of dolphins swimming underwater in a natural habitat with several kinds of fish, painted by an unknown artist at the Palace of Knossos, Crete, around 2000 BC.

The famous Dionysus Cup (540 BC) illustrates a Greek legend. A

The Dionysus Cup (Hirmer Fotoarchiv, Munich)

bearded, crowned figure, richly attired, reclines in a splendidly ornate craft shaped like a great dolphin and decorated with sea-creatures, including stylised cormorant and porpoise. The square sail is well filled. From the foot of the mast and the loins of the man two leafy vines wreathe heavenwards, bearing seven bunches of luscious grapes. There are no crew visible, but seven dolphins swim around the boat, representing the seamen who had plotted to sell their passenger into slavery, unaware that he is Dionysus, God of Wine and Frenzy (Bacchus to the Romans). Aware of their evil conspiracy Dionysus had played magic tricks on the treacherous crew. Their oars became snakes, vines sprouted from the god, flutes sounded, and the terrified sailors dived overboard.

The legend records that the drowning, repentant seamen were rescued by the Greek sea-god Poseidon (Neptune to the Romans). He turned them into dophins, the first ever known, who in gratitude for their lives became slaves to draw his sea-chariot and obey his orders. (This conveniently explained to the Greeks the dolphin's deep interest in and willingness to obey man's commands if treated kindly.) Poseidon ordered his dolphins to seek out his shy bride Amphitrite who had hidden in a cave. Later the sailor-dolphins rescued his son Taras from drowning.

The old legends are interesting because they invariably have some factual basis. Perhaps the most famous of all dolphin-human relationships occurs in the story related by Pliny the Elder, which was doubtless accepted as true by a credulous, superstitious public before the age of the sceptical modern scientist dawned and it was discredited, along with a great deal of misinformation, or misinterpretation of reported observations, which Pliny incorporated into his *Natural History*.

It will be appropriate to give Pliny's story here, as pleasantly translated in the edition by Philemon Holland published in 1601:

> In the daies of Augustus Caesar the Emperor [also the days of Jesus Christ] there was a Dolfin entered the gulfe or poole Lucrinus, which loved wondrous well a certain boy, a poore mans sonne: who used to go every day to schoole from Baianum to Puteoli, was woont also about noon-tide to stay at the waterside, and to call unto the Dolfin, *Simo*, *Simo*, and many times would give him fragments of bread, which of purpose hee ever brought with him, and by this meane allured the Dolfin ever to come ordinarily unto him at his call. I would make scruple and bash to insert this tale in my storie and to tell it out, but that Mecaenus Fabianus, Fluvius Alfius, and many others have set it down for a truth in their Chronicles.
>
> Well, in processe of time, at what houre soever of the day, this boy lured for him and called, *Simo*, were the Dolfin never so close hidden in any secret and blind corner, out he would come abroad, yea and skud amaine to this lad: and taking bread and other victuals at his hand, would gently offer his backe to mount upon, and then down went the sharp pointed prickes of his finnes, which he would

Roman public lavatory at Timgad, AD100; subdivided by arms carved as dophins, sociable latrines of this kind were a normal feature of Roman towns

> put up as it were within a sheath for fear of hurting the boy. Thus when he had him once on his back, he would carrie him over the broad arme of the sea as farre as Puteoli to schoole; and thus he continued for many yeeres together, so long as the child lived. But when the boy was falne sick and dead, yet the Dolfin gave not over his haunt, but usually came to the wonted place, and missing the lad, seemed to be heavie and mourne again, untill for verie griefe and sorrow (as it is doubtless to be presumed) he was also found dead on the shore.

There are similar stories of boy-dolphin love in later authors of the age when the Mediterranean was a clean sea lively with schools of whales, dolphins and porpoises, familiar companions of the fishermen and traders. Their engineless vessels did not disturb the cetaceans, who were rarely hunted (it was considered bad luck to kill one), and by some their appearance and behaviour were interpreted as omens of good or bad weather.

The poet Oppian (about AD 200) sang in praise of the dolphin, with rapturous words:

> On the island of Poroselene in the Aegean Sea a dolphin once loved an island boy. It ever haunted the haven where ships lay at anchor, even as if it were a townsman and refused to leave its comrade but abode there and made that its house from the time that it was little till it was a grown cub, like a little child nurtured in the ways of a boy. But when they came to the fullness of vigorous youth, then the boy excelled among the youths and the dolphin in the sea was more

A dolphin as a handle on a Chinese urn, probably the basis for the use of dolphins on European porcelain, especially eighteenth-century French work

excellent in swiftness than all others. There was a marvel strange beyond speech . . . to behold. Report stirred many to hasten to see this wondrous sight, a youth and a dolphin in comradeship, and day by day beside the shore were many gatherings of those who rushed to gaze upon this mighty marvel.

Then the youth would embark in his boat and row before the embayed haven and would call, shouting the name whereby he had named it from earliest birth. And like an arrow the dolphin when it heard the call of the boy, would swiftly speed to the beloved boat, fawning with its tail and proudly lifting up its head fain to touch the boy. And he would caress it gently with his hands, lovingly greeting his comrade who would be eager to enter into the boat beside the boy.

But when he dived lightly into the brine, it would swim near the youth, side by side, its cheek close by his and head touching head. Thou would have said that in its love the dolphin was fain to kiss and embrace the youth, so close in companionship it swam. But when he came near the shore, straightway the youth would lay his hand upon its neck and mount on its wet back. And gladly with understanding it . . . would go where the will of the youth drove it, whether over the broad sea afar he commanded it to travel or

> merely to traverse the space of the haven or to approach the land; it obeyed every behest.
>
> No colt for its rider is so tender of mouth and so obedient to the curved bit; no dog trained to the bidding of the hunter is so obedient when their master bids, to do his will willingly, as that friendly dolphin was obedient to the bidding of the youth, without yokestrap or constraining bridle. And not himself alone would it carry but would obey any other whom its master bade it carry on its back, refusing no labour in its love.
>
> Such was its friendship for the boy while he lived; but when death took him, first like one sorrowing the dolphin visited the shores in quest of the companion of its youth: you would have said that you heard the voice veritably of a mourner—such helpless grief was in it. And though they called it often, it would no more hearken to the island townsmen nor accept food when offered it; and very soon it vanished from that sea and none marked it any more . . . Doubtless sorrow for the youth that had gone killed it, and with its comrade dead it too had been fain to die.

Apart from the fact that the dorsal-fin is fixed and useful for a rider to hold on to, and that in captivity a dolphin will not eat bread (although it will courteously take many inedible objects from the hand of a beloved keeper), these two tales of the loving relationship between dolphin and human could well be true in essence. At least I am prepared, with reservation, to believe they happened, for I too have enjoyed, albeit briefly, a not dissimilar experience with a wild dolphin.

Returning in the summer of 1975 to revisit my old home of Skokholm Island off the south-west coast of Wales, I was amazed to be told by Terry Davies, the boatman who ferries visitors to the nearby island bird sanctuaries, that a large dolphin had 'fallen in love with my little brown dinghy'.

We were standing on the little pebble beach of Martinshaven, walled in by high basaltic cliffs, where Terry moors his large passenger-carrying ex-RNLI lifeboat, *Sharan*.

'We call him Bubbles. But see for yourself. Take the dinghy and splash the oars around while I am fetching the island mail and groceries from the village.' I sculled from the beach towards Terry's motorboat. Almost at once a grey-backed form poked its head above water close to the *Sharan*, then surged towards me with a half-leap—'swiftly speeding to the beloved boat', as in Oppian's tale.

The wash of his crashing dive and the bump of his arrival under this seven-foot pram dinghy nearly capsized me. I clung to the centre thwart with both hands.

Then Bubbles surfaced alongside, his mobile body turned sideways to examine me with that lively staring eye.

'He wants you to jump in and join him!' shouted Terry from the shore.

The Atlantic water here is very cold, as I found on gently stroking the back and dorsal-fin which Bubbles presented to me a few seconds later, having evidently decided I was a harmless and friendly occupant of this little brown pram, whose length of seven feet was a useful guide in measuring the size of the dolphin.

He was approximately eleven feet long, and undoubtedly a fully adult bottle-nosed dolphin *Tursiops truncatus*. But where had he come from—so tame with a complete stranger? Was he an escapee from a seaquarium, where this species is a favourite with trainers because of its intelligence and ability to live long, even to breed in captivity?

Bubbles rolled over, inviting me to do as Terry suggested, dive in and play with him. Instead I tried to take photographs, between intervals of scratching his back and belly as he continued to roll alongside or under the dinghy.

But photography was impossible, so swift were his movements which imparted a crazy motion to the tossing dinghy. Bubbles seemed to get bored, and darted seawards; he turned, making a perpendicular leap, during which—as it seemed to me—he gazed with those expressive eyes to study human movements on the beach. In fact he was assessing, I was soon to realise, the arrangements for his entertainment that morning.

For him the dinghy was a companion only at night and in the absence of human company. The little brown boat which he nudged lovingly at its moorings satisfied this sociable dolphin's need for something to possess and touch; the curving lines of the dinghy probably reminded him of some dolphin he had once known—his mother, a youthful playmate, perhaps a mate who had died? But why was he so alone—was he an outcast from a school, perhaps driven out by a rival male? There were some healed wound marks close behind one eye.

Some skindivers had just arrived and were excitedly struggling into their wetsuits. As they waded out, adjusting their snorkels, Bubbles cavorted joyously within the haven waters, making a series of high leaps, clearly in happy anticipation of what followed.

When four divers swam towards him, he winkled his way gently between them, slowly sinking to the shallow bottom. They converged on him, and through the clear water I could see him roll over for them to scratch the sensitive region of his belly.

Ever restless, he soon rose, breaching the surface between the legs of Ralph, a youth of fourteen years, who found himself astride a wild dolphin, and gasped his alarm and pleasure through his mask.

Another diver grabbed the broad flukes of Bubbles's tail. The dolphin towed the twain a dozen yards seawards then sounded—with four divers in splashing pursuit.

So the morning frolic went on. Relays of divers pursued Bubbles—some a little too roughly, I thought—whose play mood lasted until Terry Davies started the ferry-boat engine. All aboard now for my old home of Skokholm Island, piloted by the exuberant Bubbles, swerving from side to side in the surf-wave created by the bows of this fast-

The seal of Jean II, count and 'Dauphin' of Viennois in the thirteenth century

moving boat.

'And why do you suppose he does it?' asked Terry at the helm. 'No one feeds him, he gets no reward. Sometimes he goes away for days, but he always comes back. For three months now he's been with us.'

Local newspapers had featured Bubbles, and I was shown recent photographs of this very friendly dolphin, who seemed to be attached to humans generally and no one person in particular. In one account it was alleged that Bubbles preferred women skindivers, especially young girls . . .

'Absolute rubbish!' said Terry, who obviously loved his summertime companion. 'I've heard some women say Bubbles has tried to mate with them, that he's sex-starved. My opinion is that women divers are much gentler, they touch and stroke him quietly. The men are too rough—they grab at him and damage his sensitive skin. So he prefers to play with children and women.'

'It seems that what makes a dolphin seek human companionship,' I suggested, 'is probably as unconnected with sex as some human companionships may be. The famous case of Opo, the New Zealand dolphin, who befriended many bathers, had nothing to do with sexual desire, for she was just a half-grown female. She seemed hungry only for company. However, there is no doubt that adult bottle-nosed dolphins—as studied in captivity—are as well sexed as any other intelligent creatures, and will rub their genital region against objects, including a human swimmer, in their tank. In other words, like many a human being but more openly, a dolphin is able and willing to masturbate. Perfectly natural, of course.'

The mating and amatory behaviour of the cetaceans is discussed later in this book. Just here we must record the continuing history of this individual I had so unexpectedly met and caressed at Martinshaven. For it is now proved that Bubbles first appeared and made friends with humans in the Isle of Man three years earlier.

'It was on the 5th of April this year that I first saw Bubbles,' Terry told me. 'He was following my *Sharan* on the way to Skomer Island, playing with the dinghy towing behind. When we got back to Martinshaven he stayed around. He seemed to have fallen in love with the little boat, for he still plays with it a lot—but often he will push other moored dinghies around, and he likes to pull at mooring ropes and the big pink floating buoys if there is no one to play with.'

Exactly the same behaviour had been noticed in a large dolphin which had first appeared in Port St Mary, Isle of Man, in March 1972. It was very friendly until some stupid person shot at and wounded it during the summer holidays of that year. This wound, and one caused by an accident in July 1974 when Donald (as he was called in the Isle of Man) hit and bent the shaft of an outboard motor, resulted in scars by which he was identifiable. He was more wary of people and boats for a while, but continued to haunt Port St Mary. On one occasion when he was stranded on the beach at Derbyhaven by a falling tide, the residents and visitors formed a chain to keep him cool by copious applications of sea water from buckets; and a mechanical excavator was called in to dig

A dolphin decoration on the rudder of an eighteenth-century French royal launch

a canal and 'bring the sea to Donald'.

Donald was apparently greatly upset in the winter of 1974–5 by the noise of explosives used in a long period of harbour improvement at Port St Mary, and he must have left the Isle of Man about mid-March 1975. However, when the news got around that a dolphin answering his description had been living at Martinshaven since 5 April, 250 miles south of the Isle of Man, Mrs Maura Mitchell, as well as Dr Horace Dobbs, who had both swum with Donald and taken photographs at Port St Mary, visited Martinshaven and confirmed from the scars on his body that Bubbles was indeed their old friend Donald.

But his story does not end there. Donald lingered on at Martinshaven in the early winter of 1975, but by then there were few people to play with, and the dinghies and motor-boats were hauled up ashore. He must have been lonely, meeting only a few skindivers who swam out to see him on a fine weekend. He moved into the great walled-in fiord of Milford Haven, where shipping had greatly increased since the opening of the ocean terminal for gigantic tankers bringing oil to the new refineries there. He was restless and often absent for several days.

Late in January 1976 he appeared in Cornish waters, first of all at St Ives, then two days later at Mousehole. Now dubbed 'Beaky' this dolphin, which must be at least ten years old today, has remained in the area of Mousehole and the south-west coast of Cornwall, and is the subject of a continuing study by observers seconded by the Whale Research Unit, Institute of Oceanographic Sciences. Christina Lockyer (1978) has collated the information so far gathered, in her '*History and Behaviour of a solitary wild, but sociable bottlenose dolphin on the west coast of England and Wales*'.

It was a young dolphin in New Zealand which provided one of the earliest examples of a loving relationship between adolescent dolphins and human children in our century, as I shall presently relate. But a few decades earlier world-wide attention was focussed on Pelorus Jack, another New Zealand dolphin, which for many years regularly accompanied the steamer operating between Wellington, North Island, and Nelson, South Island, as it entered Pelorus Sound.

Pelorus Jack, believed at that time to be a grampus or Risso's dolphin *Grampus griseus*, is now thought to have been a large bottle-nosed dolphin. It never followed the ferry across windy Cook Strait, but would meet any large ship which entered Pelorus Sound on the southern side. Jack would follow, or more often lead ahead in the surf wave at the ship's stem, for a distance of about six miles, but was never reliably reported to pass through the tide-race of French Pass, where there is a lighthouse. Its keepers were best able to report this dolphin's behaviour.

The association began in 1888, and continued for nearly twenty-four years. Charlie Moeller, one such keeper of the French Pass Light, reported that Jack would initially come in alongside the vessel and then

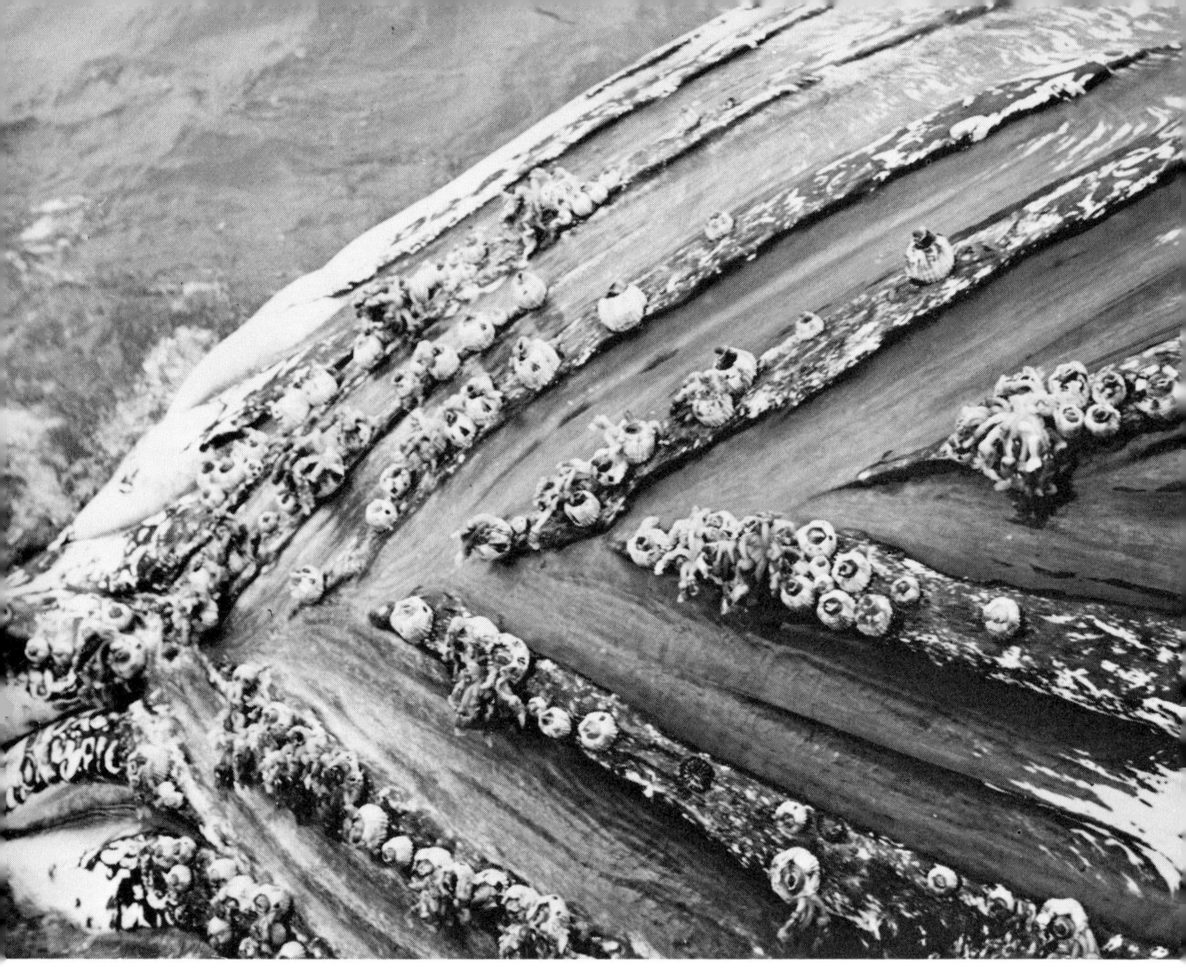

Cup barnacles on the ridges between the expansible throat/belly grooves of a baleen whale (Topham)

gradually drop astern. It was his belief that keeping pace made barnacles, which caused irritation, tear away from the dolphin's skin. Of course, other people did not share this belief, preferring to think that he was actually piloting boats safely into harbour, and many travelled from Europe and America to watch the performance.

In general the smaller fast-swimming cetaceans rarely have external parasites. The larger whales do, and I sometimes wonder if the spectacular leaping of these giants and the crashing flop upon the surface of the sea may not be partly due to irritation with, and an attempt to dislodge, the numerous barnacles, copepods, whale-lice, lampreys and other ectoparasites which fasten upon their skin.

The attachment to ships shown by Pelorus Jack is explicable today from the example of Beaky, alias Bubbles, alias Donald: the natural desire for the lone but sociable animal for company, for a substitute, lively moving, smooth-bodied object against which it can rub its own body, and enjoy the human-like sensation of being close to a loved possession—as a child loves a toy or doll. Typically, in all the cases mentioned in this book, this possessiveness was confined to the area in which the dolphin lives and feeds, that is, its territory, which it patrols, and probably enjoys with a very human-like feeling of security in a familiar, safe environment.

So regular and popular was Pelorus Jack in his assignment with the ferry-boat over those six miles of Pelorus Sound that a unique Order-in-Council was enacted by Governor-General Plunket, on 6 September 1904, prohibiting 'the taking of the fish or mammal known as Risso's dolphin in Cook Strait and the adjacent bays, sounds and estuaries, for a period of five years,' on pain of a fine of not less than five and not more than one hundred pounds. The order was twice renewed. In 1911 Pelorus Jack went missing, and was not positively identified again, despite rumours of live and dead dolphins in the sounds until late in 1912. It was popularly believed that he was shot or taken by harpoon by the crew of a Norwegian whaling fleet which anchored off Pelorus Sound about 20 April 1912. If so, the crew might be excused the killing and boiling down for oil; they probably had no knowledge of the Order-in-Council. On the other hand the Webber family farming at French Pass in 1888, and frequently using the mail steamers which Jack accompanied, considered the dolphin 'was an old fish and died from natural causes'.

The publicity achieved by Pelorus Jack ('Jack' is a term popularly applied in Australasia to any large male animal—although his sex was never known) rippled around the world, causing arguments and a flutter in scientific circles. The Linnaean Society in London investigated the facts as far as they could be ascertained. At its annual meeting in 1929, the president, Sir Sidney Harmer, author of learned papers on cetaceans, felt obliged to concede that in the light of these facts about Pelorus Jack 'we may have to review our incredulity in regard to the classical narratives of the friendliness of dolphins towards mankind'.

And so the oft-told story of Opo, which clinched the matter two and a half decades after this condescending pronouncement. Opononi Jack, as she was at first dubbed, until it was discovered that she was a female, appeared in Hokianga Harbour, Northland, New Zealand, early in 1955, near the sandy beach below the large village of Opononi. Possibly she was the well grown child of a female which had been shot earlier by a local youth. She behaved as if she was seeking for a lost parent or companion, by following small boats and nudging them, just as our Welsh dolphin Bubbles was to do twenty years later.

Soon Opo permitted the occupant of a small boat to touch her during these frolics of hers. By the Christmas–New Year summer holidays 1955–6 Opo had become so tame that she allowed numerous visitors, who swam or waded out to her in the clear shallow water of this sheltered beach just inside Hokianga entrance, to touch, scratch, even hold her briefly in their arms.

Opononi village burgeoned into a thriving small town under the pressure of thousands of curious sightseers. The local hotel and motel accommodation was booked for months ahead. The residents formed a dolphin protection committee and erected such signs as WELCOME TO OPONONI BUT DON'T TRY TO SHOOT OUR GAY GOLFIN. Lisping small children had given her this title; and it was children that she sought out and played with most, especially those who were gentle.

In particular thirteen-year-old Jill Baker, who loved swimming, was adopted by Opo, who nudged her affectionately, sliding between her legs and carrying her short distances. Later Opo allowed Jill to place little children on her back for a brief ride.

Opo quickly learned to play with a ball, flipping it high with her snout or tail. She seemed to enjoy an audience. She would 'show off', leaping exuberantly out of the water, but never to the danger of humans near her, always moving slowly in their presence. However, when annoyed by rough, clumsy hands trying to grab her, she would swim away, expressing her displeasure by repeatedly smacking her tail flat on the surface of the water.

On 8 March 1956 a special Order-in-Council gave Opo five years' protection: 'It shall not be lawful for any person to take or molest any Dolphin in Hokianga Harbour under penalty of a fine not exceeding £50.' Sadly, on the same day as the regulation became law Opo was missing. Her body was found next day, five miles up harbour, trapped in a rock crevice from which the tide had receded.

Opo was given a public funeral, and buried beside the village hall. Her beloved form is perpetuated in a statue erected close to the beach where she entertained thousands of visitors over nine months of that long summer of the dolphin.

There are several other vivid accounts in the present century of wild, chiefly bottle-nosed dolphins, coming inshore not only to play with human bathers, but on occasion to save a human life. They are said to have nudged swimmers in difficulties towards the shore, until the distressed person could feel the bottom and walk to safety. We can now accept that some of these circumstantial accounts must be true; and that others are substantially based on fact in instances where admiration and emotion have moved the author to ascribe this rescue behaviour to altruistic reasoning on the part of the dolphin.

Certainly we are encountering in the cetacean a way of life and attitude towards us that seem at times to exceed in goodness and charity those of the average human. You cannot live for long with dolphins or whales without realising that these sea-creatures sometimes appear to be happier and gentler in their social and individual behaviour than we humans are.

Often when, as a young man, I spent summers fishing for lobsters, crayfish and crabs around the tide-wracked island of Skokholm, lifting my pots, and, with no engine to deaden natural sounds, I would pause to exchange glances and unspoken words with a passing school of pilot

Boy on a dolphin, from a mid-second-century AD mosaic in the Baths of Neptune at Ostia

whales, dolphins or porpoises. I would catch that swift stare in the eye on the near side, as they breached or leapt clear of the surface. The dolphins were most playful, often returning to swim singly or in small groups around my boat. Sometimes, like seals, they would play with a rising rope or nudge at a pot coming to the surface.

But, more curious than the resident sleepy seals (which I knew individually from marking with flipper tags at their cave and beach nurseries along this wild coast), the lively cetaceans seemed to be asking questions, as if, in their fugitive glances, their circling of my boat, they were saying: 'Man, who are you? Where are you going? Are you behaving intelligently, preserving your environment, not fouling our beautiful sea as you are destroying the land we once inhabited, but long ago wisely left?'

I would wave to the friendly dolphins, and in answer receive—as I imagined—an affirmative nod, a backward flip of head or tail. At that time I knew almost nothing about the cetacean mind; I was simply amused by the friendliness expressed in their fixed smile and exhibitionism—so utterly different from the attitude of the basking sharks which came around the island in late summer. You could sail your boat bang into the soft body of this sluggish plankton-eater sleeping at the surface, without any lively reaction: it would roll ponderously away under water, its mind apparently nearly or quite blank.

About this time Frank Robson was also fishing in a small boat some twelve thousand miles away in Hawke's Bay off New Zealand's North Island. He was establishing a closer relationship with *Delphinus delphis*, the common dolphin, a world-wide species having several local types or subspecies, and often seen in huge schools.

In his book (1976) he describes how he would call them to his boat by tapping the hull with a hammer, rubbing a file on a steel rope, or blowing a dog whistle. Hitherto, this dolphin had been regarded as difficult to train and keep in captivity. But in the wild Robson was soon on intimate terms with individual dolphins, which at last permitted him to lean over the gunwale and stroke them.

He would squeak to them, imitating their squeaks. He would invite them aloud to race his boat. They invariably won. He watched them play group games near him, such as a leader picking up a piece of sea-weed as a signal for every other dolphin in the school to find another piece and hold it, or if it could not find one, to steal a piece held by a neighbour. He taught certain wild dolphins to obey simple commands, first of all by word of mouth and hand signals, and finally, as he insists, by telepathy.

His deep interest in dolphin intelligence led him to become head trainer at the new Marineland at Napier in Hawke's Bay. He devised a method of catching which was quick and caused no harm to the dolphin: a simple tail-grab with a locking device, and a long-handled rubber hoop to lift the animal's head above water immediately to enable it to breathe. The capture is completed in a few seconds after the tail is gripped; the dolphin is lifted on to a soft mattress, and kept moist under a wet blanket.

He recalls that his first capture was a female common dolphin and, as soon as she was aboard, her calf came alongside, looking for mother and squeaking anxiously . . .

> There was only one thing to be done. We took her off the stretcher and put her back into the sea and the two of them swam off to join the herd, not frightened or shocked, though perhaps a little surprised and more than a little indignant.
>
> The incident . . . taught us something. We had read that a calf, when suckling, puts its long slender tongue into the teat cleft and wraps it around the teat and the mother then ejects the milk into the calf's gullet. Our dolphin nursing mother showed us that this was wrong. When her skin was lightly touched close to the cleft, the teat was instantly ejected. When the fingers and thumb were closed round the teat and a little pressure applied, she pumped the milk into the hand. In this case the reaction may have been stimulated by the presence of the squeaking calf.

Frank Robson was engaged at Napier's Marineland at first merely as a catcher of fish needed to feed the dolphins and sea-lions. One evening he lingered to talk to the captive dolphins whose habits he knew so well at sea. To the amazement of the Marineland manager, Robson silently 'willed' the group of dolphins, watching him hopefully, to swim round the pool. They did so, and when they reassembled in front of him, he praised them aloud—and silently. The more he did so, the more they lifted head and shoulders above the surface and seemed to enjoy this admiration.

Common dolphin (Topham)

Suddenly he had the idea that in this attitude it would be easy for them to dance. In his book he says that at this point he 'concentrated my thought on them in an instruction to do the twist'. And, as he has since told me in person, they began to swing and move their bodies in time with his silent willing. So much so that he joyously copied them, and soon he, and a little later the astonished manager, were all doing the twist. The dolphins easily out-twisted the humans, having the advantage of the easier medium of water in which to indulge excited variations of these body contortions and frolics.

Robson also found the dusky dolphin *Lagenorhynchus obscurus* amenable to training, and in general a sweeter-natured creature, less aggressive than the common species. At the birth of a baby common dolphin one dusky female acted as midwife, while the father stood guard to ward off the approach of other inquisitive dolphins in the pool. When he first caught a dusky at sea, as soon as it uttered distress whistles, three companions rushed to the side of Robson's boat fearlessly, their thoughts on the need to rescue the captive being lifted aboard.

A mother dusky whose calf had died in a fishing net in Hawke's Bay would not leave her child, which drifted ashore. She remained close by, though she could hear no sound from the dead calf, and eventually, like Pliny's Simo, she lost the will to live, and died—there is no other explanation—of grief.

A skindiver at Napier who pretended to be in distress in the same pool as a dusky—as a fun act—found himself being lifted by the dolphin to the surface and held against the side of the tank. Similarly, when a common dolphin calf was ill it was a dusky who attended the dying child far more passionately than its mother, and carried it at the surface until she herself was exhausted.

Many casual observers find it difficult to believe that telepathy between man and cetacean is possible. But Robson has demonstrated its existence to the satisfaction of many who have watched him at work. Time and again, too, he has noticed that captive dolphins are aware of the silent love and sympathy extended to them by certain visitors to the pool, even though that person may be just one in a crowd. They are even made aware by telepathy, he says, of the approach of that dolphin-lover before he or she is in human sight or hearing.

Lyall Watson (1976) reminds us that our bodies emit quite large amounts of energy in the same high frequencies used by most radar transmitters. These microwaves explore everything in the vicinity so that when we are with other people we are consciously probing them, and they in turn are searching us with their transmissions. We send out extra flares of radiation when we are under strong emotion. Although 'the human body is a comparatively weak source of power, our nervous system has many of the same semiconducting properties as a transistor and can magnify weak electrical effects as much as a million times. This makes us very sensitive receivers, quite capable of picking up

Dusky dolphins (Topham)

A representation of a whale from a Japanese painting

signals from a distance. Theoretically there is no reason why we should not be able to detect messages coming from similar organisms many miles away, possibly from points beyond our visual horizon.'

If humans can be so perceptive, may not the cetaceans, with their highly developed brains, penetrating echo-location skill, and deep humane interest and response to kindness and gentleness in man, be able to perceive our thoughts, as Frank Robson and many others who have grown to love these creatures believe? I do not wish to appear sentimental about them in this book, but watching them in the wild, and seeing those marvellous underwater films of divers dancing a ballet with great wild whales and small dolphins, with close-ups of the expressive eye gazing upon the human partner, 'a great joy and a great sadness fills you when it comes to leaving these gentle loving companions', to quote Krov and Anne Menuhin (son and daughter-in-law of Yehudi Menuhin) studying southern right whales off Patagonia.

It was Pliny's contemporary, Plutarch (AD 46–120) who wrote that of all the animals, including the wild ones which avoid man, and the tame ones which are tame only because man feeds them, 'to the dolphin alone nature has given that which the best philosophers seek: friendship for no advantage. Though it has no need of help of any man, yet it is a genial friend to all, and has helped man'.

A nice thought, and one which many tales and instances like those I have related endorse. Obviously dolphins *enjoy* helping man otherwise they would not do so—in fact they will not help or obey him if he is rough and cruel towards them. It really amounts to this: that the dolphin's behaviour in helping man for no reward is not completely altruistic, not altogether a 'doing-good' act out of pure loving kindness, but because it gets a reward in its own pleasure in doing so. Even the religious monk and nun get pleasure out of devoting their lives to good works, living frugally and improving their image before God. There may never have been a disinterested action by any living creature, including

Speckled dolphin leaping twenty-five feet (Topham)

AQUATARIUM
ST PETE BCH.

man, if we care to analyse our thoughts deeply enough. Misanthropic persons get a kick out of being miserable.

Dolphins, on the other hand, have such amusing expressions, a kind of fixed smile which has often been remarked, and their sportive behaviour as they swim in the power wave before your boat, or leap clear of the sea, further endear them to us.

But, despite Plutarch's observation, I am sure that their interest in man is not altruistic. If they amuse us, we equally amuse them. They are intensely curious, but unless they can get some further amusement out of being close to us, they will soon swim away. They watch us as we watch them. Indeed, with their amazing powers of sonar surveillance which can penetrate beneath our skin, and if it it true that they can read our mind—as Robson believes—then they have the advantage of a better knowledge of us than we have of them.

It is no longer surprising therefore that the swift-thinking cetaceans make use of man whenever they find it advantageous. We can go back again to those early writers who have described instances of co-operation with man to the advantage of both.

Pliny the Younger wrote:

> The reason wherefore the dolphins be so beloved of the fishers is because they drive the fish into their nets. The fishers never do them any harm; even if they find them fast in their nets they set them at liberty. I do not mean that this is so in all seas but principally in Greece and those parts the inhabitants whereof eat no dolphins.

Apart from being sacred animals in Greece, it would be folly to kill an animal which rounded up fish for you and only fair that the fishermen should allow the dolphins to eat some of the fish, which they had driven to the nets for that very purpose.

Pliny the Elder, who was Procurator (about the year AD 70) of Provence in southern France, also wrote about this commensal activity observed in the vast salt marshes of the Camargue:

> In the region of Nimes there is a marsh, Latera, where dolphins catch fish in partnership with human fishermen. At a regular season a countless shoal of mullet rushes out of the mouth of the marsh into the sea, after watching for the turn of the tide, which makes it impossible for nets to be spread across the channel—indeed the nets would be incapable of standing the mass of the weight even if the craft of the fish did not watch for the opportunity. They make straight out into the deep water and hasten to escape the only place suitable for setting nets. When this is observed by the fishermen—and a crowd collects at the place because of their keenness for the sport—and when the entire population shouts as loud as it can from the shore, calling for "Snubnose", for the *dénouement* of the show, the dolphins quickly hear... and hasten to the spot. Their line of battle comes into view, and at once deploys in the place where

> they are to join battle; they bar the passage on the sea side and drive the frightened mullet into the shallows. Then the fishermen put their nets around them and lift them out of the water with forks. Nonetheless some frenzied mullets leap over the obstacles: but these are caught by the dolphins, which are satisfied for the time with merely having killed them, postponing a meal till victory is won. The action is hotly contested, the dolphins pressing on with the greatest bravery . . . they glide between the boats and the nets or the swimming fishermen. When the catch is completed they tear in pieces the fish they have killed.

As a soldier Pliny the Elder describes this co-operation in martial terms, but typically spoils the tale with an untruth by adding that the fishermen reward the dolphins for their help on the following day when 'aware that they have had too strenuous a task for only a single day's pay, they are given a feed of bread mash dipped in wine, in addition to the fish'.

The poet Oppian is more eloquent in describing a similar operation in Greece, but by night:

> When the fishers hasten to the toil of evening fishing, carrying the menace of fire, even the swift gleam of the brazen lantern, the dolphins attend them, speeding the slaughter of their common prey. The fishes turn in terror and seek escape, but the dolphins from the outer sea rush together upon them and frighten them and, when they would fain turn to the deep sea, they drive them forth towards the unfriendly land, leaping at them ever and again, even as dogs chasing the wild beast for the hunters and answering bark with bark. And when the fishes flee close to the land, the fishermen easily smite them with the well-pronged trident. There is no way of escape for them, but they dance about in the sea, driven by the fire and the

A Greek coin from Zankle dated 500BC showing a dolphin entering a sickle-shaped harbour

> dolphins, the kings of the sea. But when the work of capture is happily accomplished, then the dolphins draw near and ask the guerdon of their friendship, even their allotted portion of the spoil. The fishers deny them not, but gladly give them a share of their successful fishing; for if a man sins against them in his greed, no more are the dolphins his helpers in fishing.

The literature on this aspect of cetacean behaviour has grown vastly since Oppian's time. It appears that in Australia the aborigines near Amity Point in Queensland have for centuries enjoyed fishing with dolphins, to their mutual advantage. As soon as the watchful fishermen recognise a shoal of mullet approaching they rush to the water's edge and beat the surface with their spears. This signal alerts small dolphins (known as porpoises here) cruising quietly offshore to drive the mullet towards the land. The excitement mounts and soon both the natives and the dolphins are jostling each other in the shallows as they attack the mullet, the aborigines using hand-nets which they fill and pull ashore, while their allies take care of the bewildered remnant of the school.

Cousteau (1975) describes an almost identical mutual benefit association for taking mullet off the coast of Mauretania by the negroid tribe of Imragen who signal the local dolphins, by water-beating, to help drive the fish into their nets.

Freshwater dolphins like to co-operate—or should we say take advantage of human fishing activity to co-operate—in the same way. Bruce Lamb (1954), who had not then read Oppian or Pliny, relates his experience with the Indian people of the Upper Amazon district touching the boundaries of Brazil, Peru and Columbia. These natives of the Rio Tapajos protect the river dolphin *Inia geoffrensis*, the bouto as they call it, which they know as an intelligent friend able to save their lives at times. They have given it a sacred 'tapu', warning people who would kill it in order to make candles from bouto blubber that they will become blind. For when this dolphin swims near humans they can safely bathe in the river; the deadly piranhas which mass to attack a swimmer flee from the bouto, which kill and feed on these rapacious fish.

Lamb describes how at nightfall he set out with two Indians in a canoe to harpoon fish by the light of a lamp. As they paddled along, one of the Indians knocked gently against the gunwale, at the same time giving a special whistle. He said he was calling up their dolphin, who knew the signals perfectly. As soon as the lamp was lit and the harpoon ready a dolphin appeared some fifteen metres away, easily located by its 'blow' sounding every thirty seconds: 'As we progressed, the fish scattered ahead of us and went for deep water, but there they encountered our friend the dolphin, who was also fishing, and so they came rushing back to the shallows. Several times they sped back so fast they ended up flopping on the beach.'

Having made a good haul in that spot, the canoe was paddled across the river to try another place. Although they could no longer see the dolphin in the darkness or hear it breathing, it knew exactly where the canoe was headed, and as soon as fishing was resumed, the same co-operative arrangement was followed, the men driving and spearing the fish towards the dolphin, and the dolphin driving them back towards the canoe. The procedure, Lamb writes, 'differed greatly from the random feeding movements I have seen dolphins engage in on other occasions'.

River dolphins have been trained, like some sea dolphins, to live close to and obey the orders of their 'owners', who use them to drive fish towards their nets. In China *Lipotes vexillifer*, the white flag dolphin of the Upper Yangtse River, with a slender upturned snout, is or was the common property of each village or fishing group and was regarded affectionately. It too is protected by an old legend—one which tells how it saved a drowning princess.

The somewhat similar *Platanista gangetica* which inhabits the Ganges, Bramaputra and Indus rivers has been trained from time immemorial to herd fish shoals towards the nets of fishermen who regard them as sacred creatures in these holy rivers. It is so tame that it swims freely between the legs of bathing and wading pilgrims who wash in the sacred waters below the shrines.

Living in the sea: Evolution and anatomy

I will not conceal his parts, nor his power, nor his comely proportion. Upon earth there is not his like, who is made withoutfear.

The Book of Job

It is believed that life began in the sea, in the warm turbid oceans of Earth's childhood. Hydrogen, carbon and other elements combined and regrouped in ever-changing formulae. At last air and water were cool enough for organic life to evolve. The protozoan cell appeared, the novice tissue with a tough but permeable membrane which assimilated nourishment under the strong sunlight and multiplied by division. The first simple plants arose—primitive phyto-plankton composed from these chemicals, energised by the benediction of light and heat, producing chlorophyll, starch and sugar.

Some plants remained floating in the sea, unchanged down a million epochs. But others developed into multicellular structures, with increasingly complicated lives. Some were successful; their descendants are still living in the sea: sponges, worms, molluscs, arthropods and crustaceans. Others died out, to became present-day fossils.

The first vertebrates appeared some 500 million years ago. They swam in the sea as reptiles and fish; gradually some took to the land, and some to fly in the air. The old shifting continents were invaded by a vast number of air-breathing animals which crawled on four or more legs and multiplied to such an extent that some were under pressure to return to the sea. Among these the cetaceans, after a long sojourn as four-footed, hair-covered creatures on dry land and in rivers, returned first to the edge of the water, and finally became wholly dependent on the sea, while still retaining modified air-breathing lungs.

Earlier attempts, from Darwin onwards, to trace the evolution of the cetaceans and some other mammals, including man, by devising a family tree (lines of phylogenetic descent) from common structural and other physical affinities, have lately been criticised. It is realised increasingly that there is a confusing parallelism and convergence, as well as divergence, in widely unrelated groups of animals, which have adopted a similar way of life, or ecology, in a similar habitat.

This is vividly evident in the rapid speciation which occurs in insular isolation. A classical example is the pouched (marsupial) fauna of Australasia, where the whole range of native terrestrial mammals (including rat, mouse, squirrel, wombat and kangaroo) is unique in giving birth to tiny legless embryos, blind, naked and worm-like, which yet are instinctively able to claw their way with hooked forearms from the vagina up into the maternal pouch, in which warm haven they are suckled and remain over a long period, gradually acquiring hind legs and a coat of fur. The marsupial baby remains suckling in the pouch for approximately the equivalent in time to the combined gestation and lactation periods of an outwardly similar range of pouchless placental animals of much the same size (including mouse, squirrel, badger, deer) which inhabit the continental mass of Eurasia-America.

Both pouched and placental have similar vertebrae, internal organs, hair, claws and body structure, showing they have sprung from a common viviparous ancestor which nourished its young with maternal milk. But evidently the first ground-living mammal to reach the island of Australia must have retained the pouch as an equally successful method of carrying and protecting its young. And so from those first single colonists arose the present range of pouched native mammalian fauna of Australia in all its numerous orders and families: surely the world's most remarkable and fascinating instance of wholesale parallelism—plagiarism as one scientist has dubbed it—or you can call it convergent evolution. The pouched descendants of the original colonist simply occupied those same ecological niches in the new land which pouchless rat, mouse, squirrel etc, occupied in the older ancestral home beyond the sea. In doing so they evolved into similar forms in order to exploit the environment, to survive and thrive, as most of them do today.

Convergent and divergent changes in the evolution of the grand order of whales, whereby some became the largest mammals ever known, are less easy to trace, and there are many missing links. But it is clear that

Remains of the head and one half of a jawbone of a baleen whale taken in the English Channel about a hundred years ago (Topham)

the cetaceans share their remote ancestry with land-based marsupials and all other warm-blooded, air-breathing mammalia (having mammae to suckle their young).

The cetacean foetus reflects this evolutionary history, showing a typical terrestrial mammal form with four limb-buds, a pelvis and tail. Each forelimb has five fingers. Although the baleen whales are toothless in adult life, they have teeth in the early embryo stage, which are absorbed as this sub-order develops the curious plates in the huge upper jaw, typical of these plankton and shrimp swallowers.

Tests of the blood proteins of modern sea-living whales, have indicated a similar composition to those of land-living deer, cattle, pig and other two-toed ungulates, which are believed to be descendants of a primitive small shrew-like ancestor which lived 120 million years ago, and is generally considered to be one of the earliest of true mammals to walk on dry land with four legs.

One branch of the shrew's descendants remained on the land and developed cloven hoofs—the horse-like artiodactyls. Another branch went back to the sea, river, lake, eventually giving rise to now extinct creatures resembling both crocodile and a lean serpentine whale.

The earliest known fossils resembling the living whale were first discovered in North America early in the last century, and subsequently in Europe, Africa, New Zealand and Antarctica. Known as *Archaeoceti*, they lived in the sea some 45 million years ago. Some of these fossil skeletons are short, close to living dolphin species in shape; others are longer and lizard-like. The forelimbs are well developed, with fingers; the arms and elbows were less shortened than those of cetaceans today. In Zeuglodon the nasal openings have not migrated far from the snout, but in Agorophius they lie somewhat above the large eye openings. All had formidable flesh-eating type teeth.

About 30 million years ago the first fossil *Squalodontii* cetaceans appear, not unlike the living river dolphins. However, all these fossil forms died out about 25 million years ago leaving no clear link with living species. One difference lies in the much larger brain in modern cetaceans, to accommodate which the skull is by comparison a massive telescoped dome, a thinker's brow. And we have good evidence that at least 10 million years before man's early hominid ancestors developed a thinking brain beyond the ape stage, the cetacean was already well-endowed with a highly perceptive brain: probably the most intelligent evolved at that time.

Perhaps the missing link, the intermediary ancestor between the *Archaeoceti* and the modern cetaceans, will be found one day in fossil form where it perished, in the deep ocean or the mud of deltas not yet examined. The baleen whales *Mysticeti* are said to be closest to the extinct *Archaeoceti*, from the rather slender evidence of skull resemblances. They too branched off into the present three families of right, fin and gray whales as they became more efficient in exploiting the

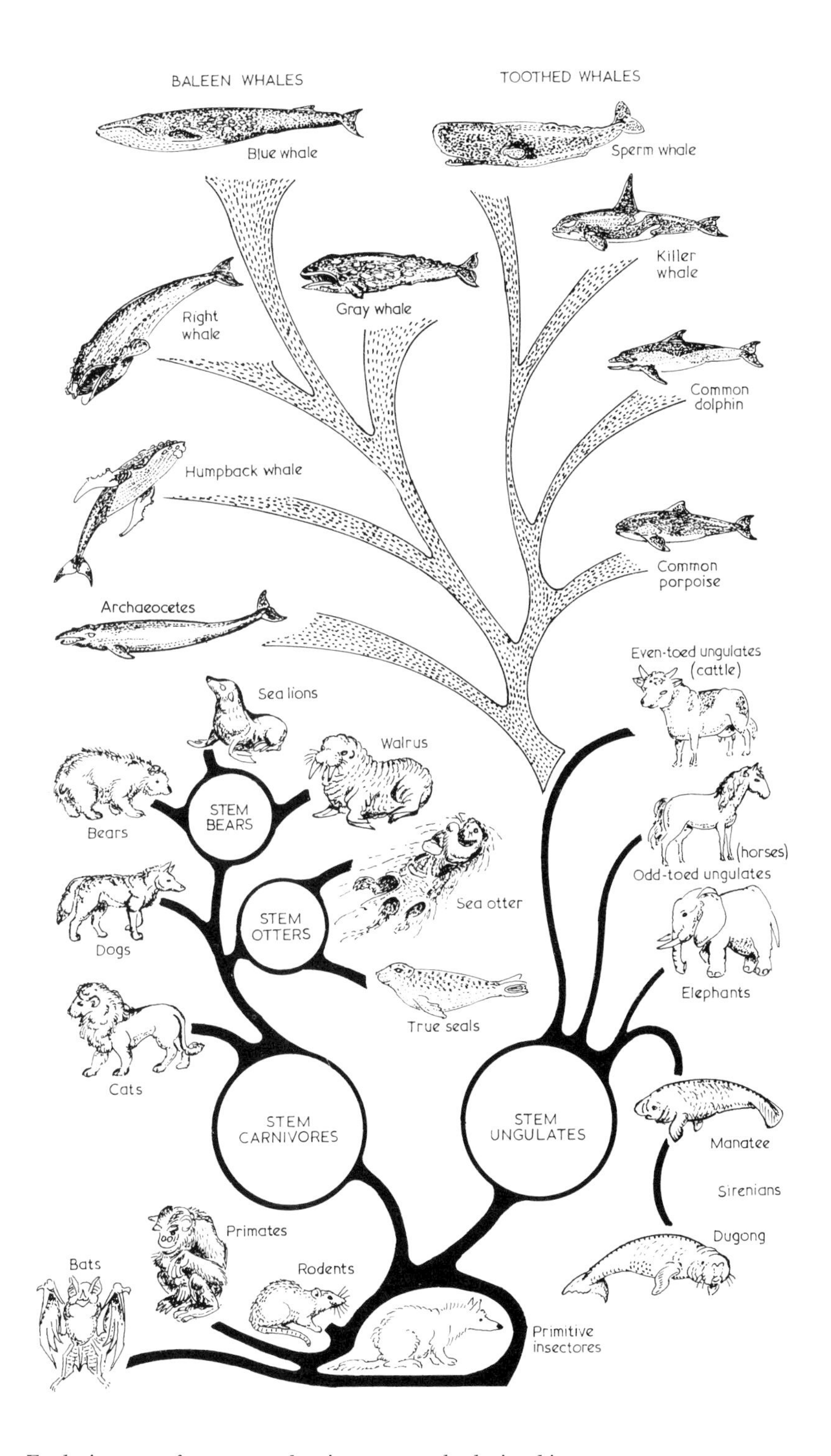

Evolution tree of cetaceans showing suggested relationships

shrimp, plankton and little fishes of their modern diet.

The more numerous toothed whales and dophins, the *Odontoceti*, evolving about the same time or earlier, divided rapidly into their present families and genera as sixty-eight species to occupy the fish-feeding niches of the oceans of the world. Some did not go far from the great rivers, which are likely to have been the site of their first return from land to sea; four riverine dolphins have remained somewhat isolated, and are sometimes referred to as primitive in early descriptions of their structure and behaviour, because they are myopic or blind and dwell in turbid water; but we know that they are just as intelligent, and expert at direction-finding, as any clear-eyed sea dolphin.

Some of these toothed cetaceans have become sedentary, living in small isolated groups in the open ocean, a state favouring rapid sub-speciation, leading to new speciation where a species is few in number and adapting to a vacant habitat geographically and genetically separated from the parent stock. Some of the smaller non-migratory species appear to be very rare, not because they were exploited by man, but because they have seldom been seen. Several are known only from rare strandings of single individuals.

Anatomy and physiology

The cetacean brain has a complex cortical area, with a greatly convoluted appearance, indicating a highly specialised 'thinking' potential. The central nervous system is intricately developed and its mechanism not fully understood. The supralimbic area, each side of the rhynic cleft, which monitors memory and conceptual thought, including sociability or association (love of community and children) is very large. The neural connections to the senses, especially of hearing and voice, including the curious 'melon'—the enlargement of the forehead—are very different from those in man and are part of cetacean specialisation as a marine mammal.

So much sympathy and emotion has lately been lavished on the large whales because they have been hunted to near-extinction, that it is difficult to be purely objective in trying to measure cetacean intelligence in terms of human thinking capacity. Their IQ has been compared favourably with that of man; it is generally agreed today that it exceeds that of the higher non-human animals such as the trained dog, sea-lion, otter and even the ape. In fact the cetacean brain has a higher neocortical-limbic ratio than the average intelligent human. As studied in captivity, several kinds of dolphins, including the orca or killer whale, have shown human-like powers of sympathy, empathy, humour, sulkiness, toleration (of other species and man) and especially self-control and memory. They love to tease and play tricks on each other and on humans close to their tank; they are intensely curious; they can be jealous, yet are normally

Dusky dolphin breaching (Jen & Des Bartlett/Bruce Coleman)

gentle and affectionate. They display almost every human emotion you can name—in short, cetaceans are almost human.

We must leave this appraisement for the present and consider the senses, which are monitored by that complicated convoluted brain.

Supported without effort in the dense medium of water of equal pressure around it, the cetacean has not the same difficulty of equilibrium as the land animal. The amount and even distribution of buoyant blubber fat, of oil in its bones, of body fluid, of air in its lungs, balance the heavier-than-sea-water skeleton bones and protein so that it floats at ease on the surface, on the level keel of its boat-shaped body, its

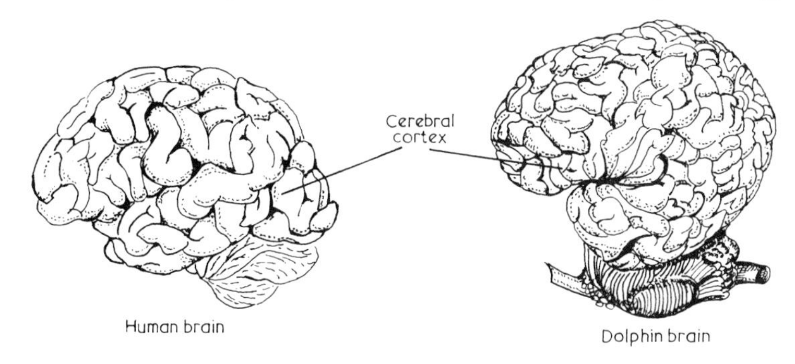

(above) *Comparison of human and dolphin brains. The human brain is highly convoluted and has well defined layers; the dolphin brain compares with the human brain for size and is highly convoluted, but not so well defined. It is difficult to compare the functions*

(below) *Skeleton of right whale and skull of bottle-nosed dolphin*

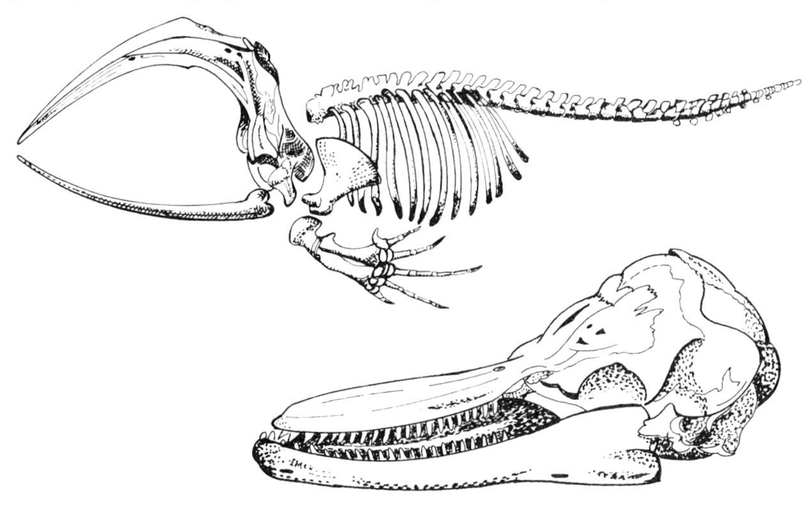

Some of the 253 false killer whales stranded near Auckland, New Zealand, in 1978 (Ronald M. Lockley)

A beluga or white whale underwater (Norman Tomalin/Bruce Coleman)

Killer whale (Pat Morris/Ardea Photographics)

flippers used as stabilisers or lee-boards. Its sense of balance resides in the semi-circular canals of its ears as in other mammals.

In captivity it is seen to regulate its swimming routine according to the feeding and training visits of its keeper. In the absence of human spectators, and after an hour or so of activity, it will generally rest, often catnapping at the bottom of the tank, but only for short periods, when it may close both or only one eye. It needs to breathe at the surface about twice a minute during daylight hours, but sometimes sleeps longer, perhaps for a few minutes at the bottom. Normally the soundest sleep occurs as it rests by night four-fifths submerged; the top of the head is then lifted clear automatically to expose the blow-hole in a breathing rhythm of two or three brief exhalation-inhalation snorts, each lasting about one second, once or twice each minute; one or both eyes are closed and the body floats almost motionless, the flippers and tail dangling inert.

Touch

The sense of touch is as well developed as in man, as anyone will realise who has lightly put one finger upon a dolphin or small whale at rest, or with its eyes shut. Perhaps the huge, slow-moving baleen whales may be less sensitive to touch (one hopes) because they suffer much from a thick encrustation of external parasites and commensals attached to the skin, and thickest on the underside: two or three species of barnacle (sea-

Humpback whale underwater showing eye, throat grooves, white under flipper and barnacles and weeds attached to throat and flippers (Al Giddings, Survival Anglia)

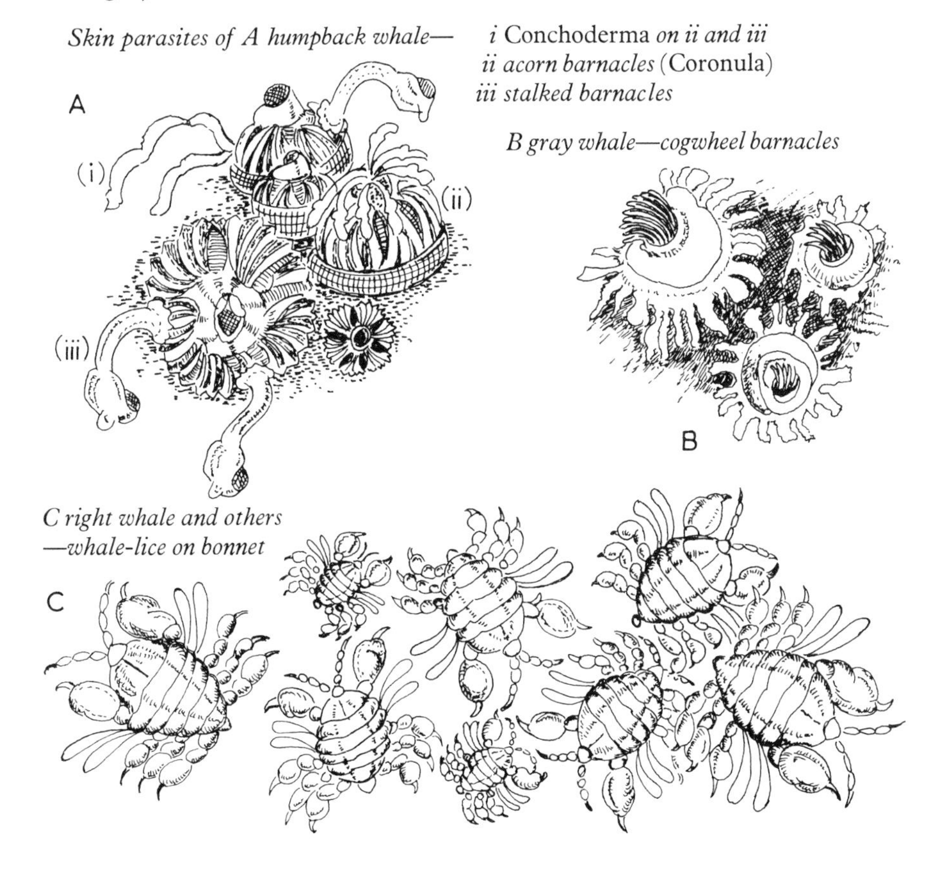

Skin parasites of A humpback whale— *i* Conchoderma *on ii and iii*
ii acorn barnacles (Coronula)
iii stalked barnacles

B gray whale—cogwheel barnacles

C right whale and others —whale-lice on bonnet

weeds are often attached to these) feeding not on the whale but on drifting plankton; remora (whale-suckers) which bite, and lampreys which actually suck blood and secretions as they fasten securely for long periods at one spot; numbers of crawling whale-lice, and often a diatomous or algal film. Many of these parasites drop off when the migratory great whales reach waters too cold or too warm for these uninvited travellers. The skin will heal, but the scars of attachment remain as patches of a different colour.

The irritation from this source causes the sufferer to rub the afflicted area against submerged objects, such as the sea floor, and occasionally against floating objects, such as the keel of a ship—an alarming experience for the occupants when the whale is large and the vessel small.

The great whales habitually leap clear of the surface of the sea, in play, courtship and high spirits; and it may well be that they are also trying to shake off these ectoparasites in the violent belly- or back-flop as the great bulk of many tons crashes flatly back upon the surface. The very long flippers of the humpback hit the sea resoundingly, an action which helps to dislodge dead barnacles. The smaller fast-swimming dolphins and porpoises, much less troubled by such annoyances, are seen to rub their lively bodies against boat keels and underwater rocks.

The extremities of the body—snout, flippers, back-fin and tail-flukes—are freely used in sexual and other tactile relationships. Mated pairs or friends will swim so close that their fins frequently touch. The child almost rides on the parental back or touching its side. The dolphin snout is used to nudge and investigate any part of another body or object. It is pushed affectionately against a young dolphin or into the genital slit of the opposite sex; or the genital region may be rubbed against fixed or floating objects, including any protruding portion, such

Humpback whale breaching

Stranded white-sided dolphin (Bobby Tulloch)

as the fin or tail-fluke, of the body of a companion, in a masturbatory action.

The main difference between the 'locomotor' systems of the cetacean and the land mammal is the changed use of the forelimbs. Except in the humpback whale these are short flippers which regulate stability and steering, rather than assist fast propulsion. The tail is the main and very effective swimming machine, levering the cetacean forward by a vertical curving pressure, that is, up and down, not by a side-to-side movement as in fish. This 'tail' has no bones of the hind limbs incorporated, as sea-lion and walrus tails have, but is strengthened by a flexible extension of the last vertebrae of the long backbone, beyond which the two flukes, joined but marked by a median notch, are boneless flesh and muscle. The hind legs with which the ancestors of the cetacean used to walk on land, and which appear as limb-buds in the young foetus, completely disappear before birth; in the adult skeleton the only sign is an unattached vestigial pelvic bone (this is absent from most species) to which the pubic muscles are anchored.

The majority of cetaceans have a well defined fixed dorsal or back-fin. It is longest in the killer whale, less prominent in the slow-moving baleen whale, almost a pimple in the humpback whale, and missing in the black finless porpoise, right whale dolphin, Greenland, gray and black right whales. This absence of a dorsal fin which it is assumed has a stabilising effect is not easily explained. It may be compensated for by the longer flippers in the finless cetaceans, which are usually fast swimmers.

The flippers can be moved independently. They assist the tail in rolling and changes of body position, as in play, feeding and diving activities. When swimming at high speed they are normally trailed back

Comparison of dorsal-fins: (above) lesser rorqual or minke whale, (centre) common porpoise, (below) Risso's dolphin (Bobby Tulloch)

against the body (as in seals and penguins), only the tips projecting, serving in that position as small stabilisers.

Unlike the flippers, the dorsal-fin has no bone attachment to the skeleton, but is strengthened by subcutaneous fibres extending from the broad back. Although one function must be tactile use in informing the owner of its position at or close to the surface of the water, this fin also serves to regulate body temperature by counter-current heat exchange. The lively, toothed whales and dolphins which have the largest dorsal-fins might become overheated in their thick coats of blubber as they swim actively near the surface if the increased flow of blood from the heart was not rapidly cooled in the epidermal layers of the dorsal-fin, as well as of the flippers and tail, and through the veins and arteries close to the surface of the rest of the body.

We have already described how the stranded cetacean suffers from this overheating. The skin blisters and the cetacean will die in distress unless it is kept cool by libations of cold water. The cetacean has no sweat glands so cannot cool itself by the evaporation of body water in air as man does. It loses water only through urine and faeces; and, as in seals, a minute constant trickle of oily fluid lubricates and protects the eye from direct contact with water.

Diving

Normally the sea constantly cools the cetacean, keeping its skin temperature close to that of the surrounding water. When it dives deep into the colder layers towards the ocean abyss its circulation is automatically reduced in proportion to the depth. Blood flows more slowly through the sea-cooled peripheral veins and arteries, thus conserving the internal heat. However, a flow of warm blood is essential to maintain the brain at full consciousness; this is achieved by a process not fully understood. Although the brain and its blood vessels and nerves are maintained at normal working temperature, the heartbeat drops during a long dive to about half the normal rate at the surface. In the killer whale the reduction is from 60 to 30 beats per minute; in Dall's porpoise and some other small dolphins from 120 to 15. In some seals (also beaver and hippopotamus) the heartbeat rate falls surprisingly low—to about 10 per cent of that when breathing at the surface.

Heartbeat rate in all these deep-diving mammals increases rapidly during breathing (sometimes called panting) spells after diving, when oxygen is quickly assimilated by the blood through the lungs. The rate decreases again during normal breath-holding pauses at the surface, until it reaches and stabilises at the normal rate, close to that of man's heartbeat.

In diving, the ambient pressure of water upon the body increases in proportion to the depth. The cetacean compensates for this by special adaptations, which include a flexible, muscular rib-cage which yields but will not break or crush the vital organs. Also, the complicated lungs

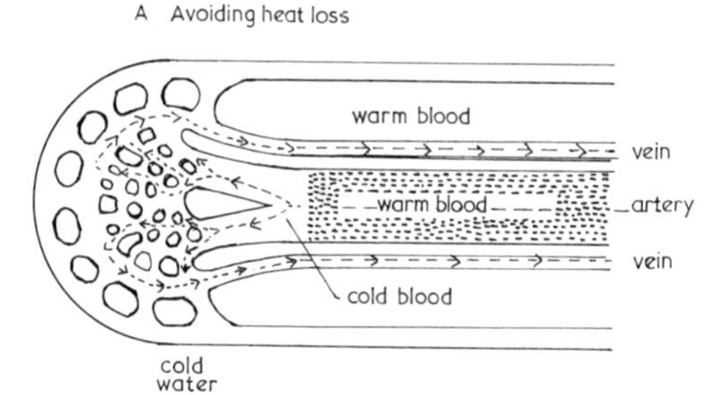

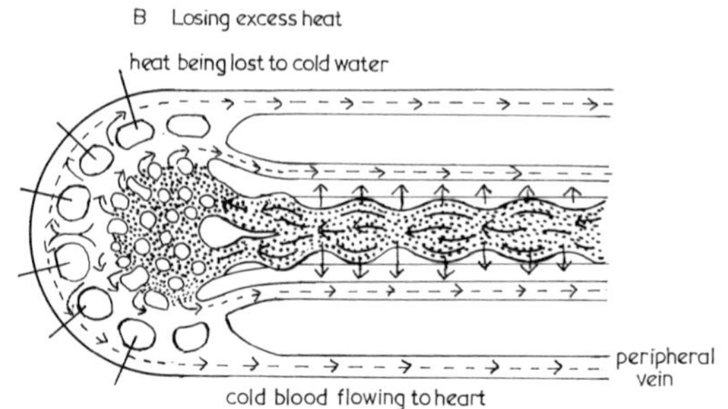

Circulation—a theory of the heat-exchange system in flippers, tail-flukes and dorsal-fin

A The arteries leading from the body to cooler extremities are surrounded by veins from the same areas; this warms the blood returning to the body and heat loss is minimised

B When the animal is overheated the blood pressure rises causing the central arteries to the extremities to expand with increased blood supply. Because the veins surrounding the arteries cannot carry all the blood back to the body, some blood returns via peripheral veins making it nearly as cold as the water

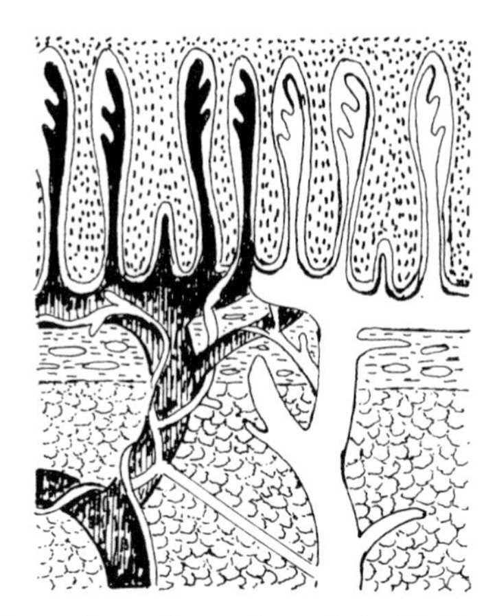

Circulation—the blood supply to the surface of the blubber and dermal papillae of a porpoise. After Parry

are composed of spongy elastic tissue and cartilage, and a series of air pockets guarded by sphincter muscles or valves.

Contrary to popular belief whales and seals do not overfill their lungs with air before diving. This would make them too buoyant. Depending on the depth of the dive they breathe rapidly at the surface, renewing the oxygen in the blood stream, and some new oxygen is trapped in the many small vascular pockets in the lungs and connecting passages. Returning to the surface after the dive, the foul air is exhaled explosively as the thick fibrous blow-hole plug or valve opens by a muscular effort. In the large whales a visible 'spout' is created by the exhalation of moisture-laden mucus along with the used air and some sea water trapped outside the blow-hole. Inhalation follows instantly, and the nasal trap-door is then slammed shut as the whale relaxes the muscular hinge back to its normal resting position. Observing the individuals of a school of false killer whales which stranded on a beach in New Zealand, I found the

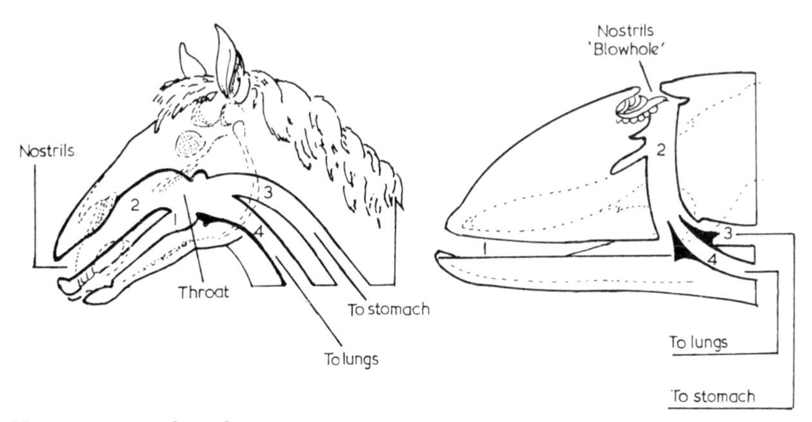

How a cetacean breathes
1 Mouth — *3 Oesophagus*
2 Nasal passage — *4 Windpipe*

intervals between each single loud blow (the nasal plug opened for approximately one second each time) were regular at 20 seconds apart, as in captive dolphins behaving normally at rest near the surface of their tank. Long immersion is followed by long panting at the surface. Whale-hunters have a saying that for every minute a whale remains below it must spout once; and this is generally true.

The largest whales, which sometimes descend to over 1,000 metres (3,300ft) in order to travel so far need to remain below for long periods of up to an hour, but generally much less. It is near freezing at that depth in temperate and polar seas, but these giants keep warm internally because the enormous bulk of the body loses the inner metabolic heat slowly through the proportionately small area of the body surface exposed to the chill water.

These deep dives seem not to incommode whales at all, except that they 'pant' at the surface for longer periods afterwards. The explanation seems to lie in their blood having a low intake of the stored oxygen during the reduced heartbeat; the brain and nervous musculature are better able to resist the invasion of carbon dioxide and lactic acid than in human divers. When whale, seal or man surfaces rapidly after a long deep dive, which exerts enough pressure on the body to force air from the lungs into the tissues, that air, now become more nitrogenous, expands with the lessened pressure, and in man produces those air bubbles in the blood which results in caisson sickness—'the bends'.

The network of numerous large and small blood vessels, veins and

Finback whale inhaling (Topham)

arteries known as the 'wonder-works' (*retia mirabilia*) found in mammals generally, is particularly large and well cushioned in fat in the cetacean. These vessels are linked to the brain, spinal chord and thorax, and seem to function as elastic reservoirs which accommodate and control the flow of arterial blood to the brain during the steep dive and that swift rise to the surface afterwards which causes the dangerous sickness in man unless he decompresses in a special chamber.

Deepest diver is the sperm whale, which has been recorded by whale-hunters as taking 1,500m (5,000ft) of harpoon line perpendicularly downwards. More accurate is the record of a sperm whale found tangled in a telegraph cable lying on the floor of the sea at a depth of 988m (3,240ft). Evidently this individual had grabbed the cable in the under-water darkness in mistake for the arm of a large cuttlefish, its chief food: in its efforts to free itself, it had wrapped the cable around flipper and tail while still gripping it with its powerful jaws. From this rare happening we can imagine the epic fight which takes place when this whale, scouring the sea bed with its toothed lower jaw at the ready, and locating a giant octopus or squid with its sonar scanner, grabs the cephlapod. The great head region of the sperm is usually marked by the suckers of its prey, evidence of this struggle imaginatively described in true and

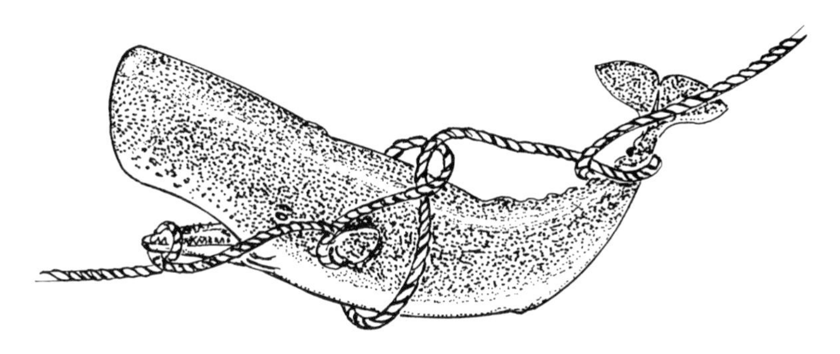

Sperm whale tangled in deep-sea submarine cable

fictional stories of whale-hunts (eg *Moby Dick* and *The Cruise of the Cachalot*).

Another sperm whale was trapped by a cable at 1,145m (3,755ft). Modern asdic locating of diving whales indicates that the cachalot regularly dives to feed at the bottom of the sea at depths of 200–1,000m (650–3,300ft).

Blubber

A healthy cetacean accumulates a layer of fat many centimetres thick known as the hypodermis, beneath the much thinner (usually less than 1cm) outer skin or epidermis. It was at one time considered that the

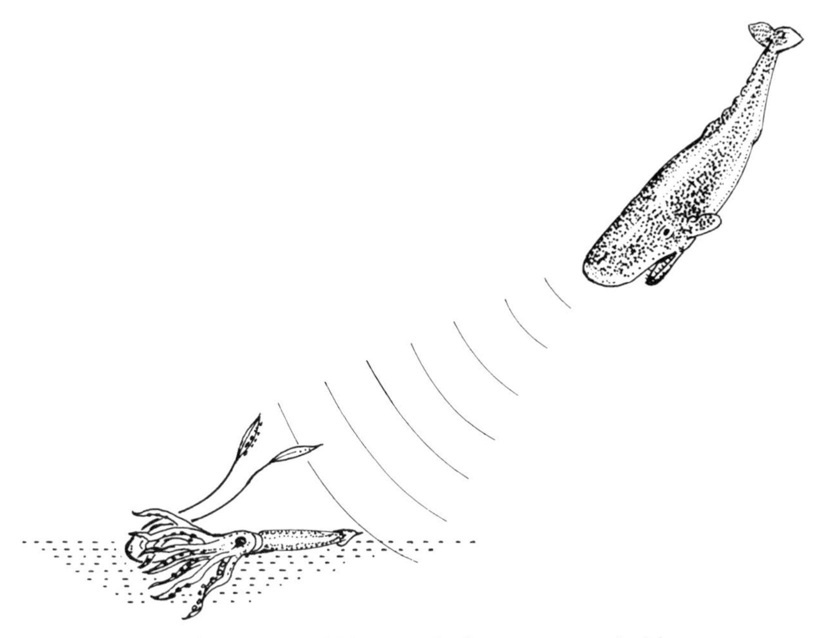

Sperm whale hunting giant squid in near-darkness on ocean bed by sonar (echo-location)

chief purpose of the blubber layer was to insulate against heat loss, which it must do in cold seas, the main feeding grounds of the great blubber-producing whales. But it is also a considerable fat (energy) reserve to be absorbed during the long fasting periods of their migration to, and sojourn in, subtropical waters where, it is said, most baleen whales feed little or not at all, although they indulge in quite strenuous courtship and mating activities, and the calves are born there. The calf at birth is without a layer of fat, and would probably die of cold if born in those sub-polar seas. It will migrate with its dam after it has acquired enough hypodermal fat during the early weeks of suckling the rich milk produced copiously (and surprisingly) during the alleged winter fast of the mother.

Whales, fat and in peak condition, moving from cool and polar seas in the autumn towards their winter quarters, might suffer overheating problems on entering warm water, but they appear to migrate at a leisurely pace, perhaps instinctively aware that the sea temperature gradient change must be controlled. If they are too warm at the surface, they are able to cool off in the colder layers below. The stomachs of those killed by whalers during their autumn migration are usually empty. By the time they reach winter quarters in subtropical seas they will have lost much weight.

The outer skin, which blisters so easily if allowed to become dry in sun and wind, is attached flexibly to the blubber layer with pimple-like papillae which, although holding the two close together, gives the outer skin enough mobility to be able to wrinkle into small ridges under water pressure, with a wave-like rippling which offsets the effect of turbulence—typical of water moving against the smooth surface of boat hulls, or of air against the sides of aircraft. To some extent the long hair of otters and fast-swimming seals, and the close oily plumage of the swift-diving penguins, may serve a similar function of reducing friction.

Visual system

The cetacean eye is normally as flexible as that of man. The eye ball moves freely to look ahead as well as sideways. It has strong upper and lower eyelids which protect the eye in deep water. The eyes are far apart, separated, but also partly protected during fast swimming by the massive head and 'brows'. It is not possible for the large whales to look straight ahead or astern with binocular vision. The visual field of a humpback whale was found to be 175° each side, nearly 360°: that of the huge-headed cachalot must be much less. Whale-hunters liked to approach silently from the rear, and keep within the blind spot, the arc of view obstructed by the body. The smaller dolphins and porpoises are too lively to be surprised in this way.

Long sight under water is limited to a dozen metres or so, even in clear conditions. In captivity, as well as in the wild, cetaceans 'watch' each other and the movements of other animals under water; but with-

Close-up of white-sided dolphin's head (Bobby Tulloch)

out eyelashes or brow hair, the region of their eyes does not serve as a signal to express emotion to the extent it does in man. But during moments of great distress or anxiety, the eyes open much wider, as in man. The often-remarked 'smile' of the dolphins is a figment of man's imagination: it is a fixed feature of their physiognomy. The heavy brow does not frown as it is too rigidly attached to the skull, as the skull is to the vertebrae (save in some river dolphins).

In the toothed whales, however, the mouth is freely opened to expose the teeth in expressive gestures which range from pleasure to anger and threat signals. The back region, behind the dorsal-fin and ribs, is comparatively supple, and freely arches in preparing to dive, in play, or in an aggressive attitude with the head bent down towards the object of displeasure. Joy, and contented moods, even in the enormous baleen whales, are expressed by various rolling movements of the body, turning and twisting, and leaping. In rolling, the genital region is deliberately presented to a companion; this action is a natural part of courtship and sexual solicitation.

Sight, however, is not absolutely essential to the survival of many marine animals, including cetaceans, seals and some fishes which have large eyes. I have several times encountered adult bull or cow Atlantic seals long blind from some cause, but still perfectly healthy and fat, although they could not see, but only hear, my approach. Two blind mother seals, one in Wales and one in the Orkneys, successfully reared a calf two seasons in succession, each time 'feeling' the way home to their beach nurseries by kinaesthetic memory.

In captivity a dolphin, fitted with suction cups which totally obscured vision, located food fish at random in the tank by other senses—chiefly echo-location. More astonishing, blindfolded dolphins will leap through hoops held above the water—provided they have become familiar with this trick by previous training without blinkers. This also appears to be a feat of kinaesthetic or muscle memory.

Adapting to life under water, often murky and at great depths where sight is partly or wholly useless, the cetacean has developed voice and hearing to such a degree that it literally sees with these senses.

Hearing and voice

The acoustic system is the most important cetacean equipment for survival, uniquely developed for locating its food, direction, companions and enemies. Hearing is acute: while man can hear sounds between 30 and 18,000Hz the cetacean range is 16 to 180,000Hz.

As land mammals originating in the sea, hearing evidently evolved in them from an aquatic ear. But if their later terrestrial ancestors ever had, like man or rat, an external ear to funnel sound to the inner ear and brain, it has long been streamlined out of existence as an encumbrance to swimming under water.

Today it is sometimes difficult to locate even the site of the ear, which is reduced to a tiny hole a few millimetres wide, just behind the eye. In the baleen whales this pinhole is blocked internally by the tip of a horny plug which fills the meatus (duct), and is up to 1.2m (4ft) long and S-shaped. Such a plug has been used to assess age, since one—possibly two—distinctive rings or laminations are laid down annually.

However, the age of young baleen whales, up to maturity, can be reliably determined by growth lines in the form of transverse ridges on the baleen plates. As in tree growth there is a distinction between the wider rings of summer growth and the narrower ones of winter, the latter appearing as a depression between the former. This ageing method, in conjunction with an examination of the reproductive organs, has proved that sexual maturity in these giants is not reached before five or six years. After this age growth rings on the plates are not a satisfactory guide to age, as the tips of the baleen begin to wear down by constant use as fast as they grow.

The elongated ear-plug embraces at its inner end a glove or finger covering the outward-projecting cone of the ear-drum of the middle ear. Some physiologists believe it is not a very effective conductor of sound and could be an atrophied relic of the land mammal ear. It has only a very narrow duct in the centre which conveys sound vibration, probably of low frequencies, to the ear-drum. Nevertheless these whale-bone whales are good vocalists, and with ears so far apart on the enormous head they hear each other directionally over considerable distances under water where sound travels much faster and farther than in air. It has been estimated that they can utter, and register, low frequency calls (16 to 400Hz) over a distance of 160km (100 miles).

There is no plug blocking the ear of the toothed cetaceans: in these a narrow winding channel admits water as far as the ear-drum, which is a cone projecting into the middle ear. Inside this drum, as in land mammals, is a space containing the auditory ossicles. Whales, however, have no eustachian tube filled with air to equalise external pressure on the drum, no epiglottis at the back of the tongue conducting air between the lungs and mouth. The mouth is solely concerned with eating and grasping, and without air can produce no vocal sound. Instead, the air space

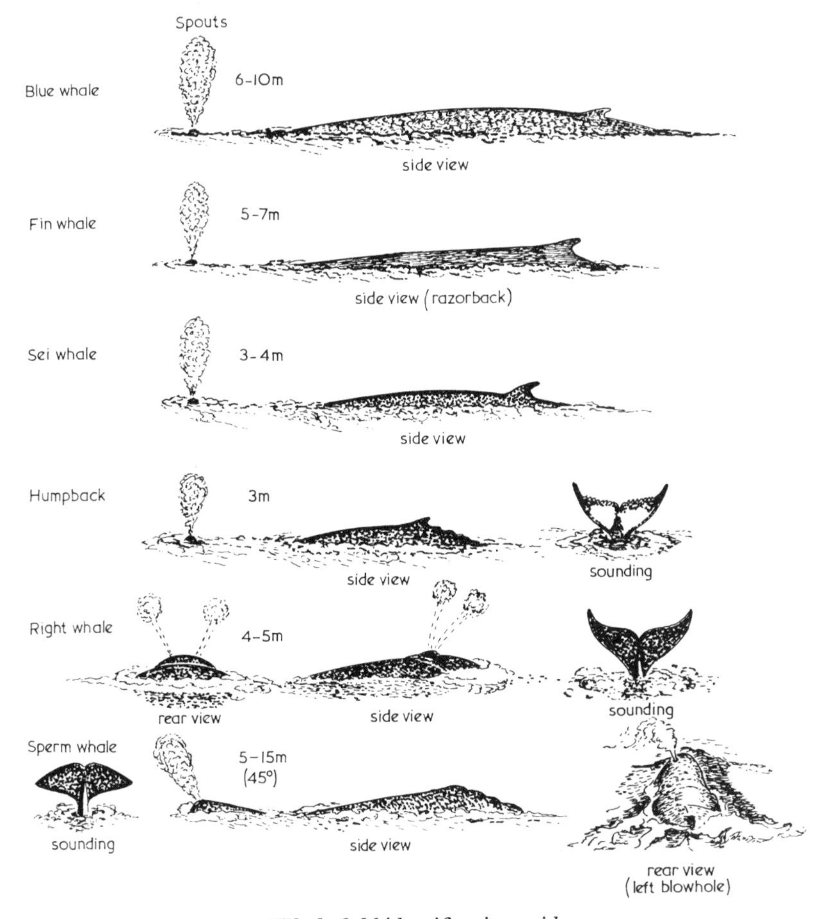

Whale field identification guide

behind the drum is surrounded with communicating cavities filled with a mucus foam secreted by the walls. This foam-filled air acts as a pressure equaliser during diving. Some of the mucus, together with the water caught externally in the depression of the blow-hole(s), is voided with the spent air from the lungs when the whale surfaces to breathe, and in the large whales appears as an explosive spout of varying height, visible and often heard for a mile or more in clear weather. The baleen whales have two blow-holes at the surface; in the toothed cetaceans the nasal passages are joined below one blow-hole to make a single spout. 'By their spouts you may know them' is a true whale-watcher's saying. The

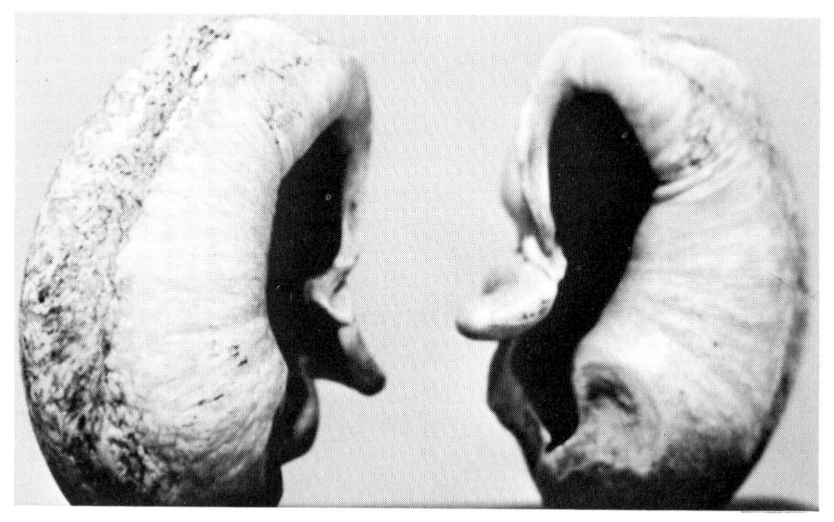

Hearing in whales; cavity view (above) and back view (below) of bulla or tympanic bones from the ear of a great whale (Ronald M. Lockley)

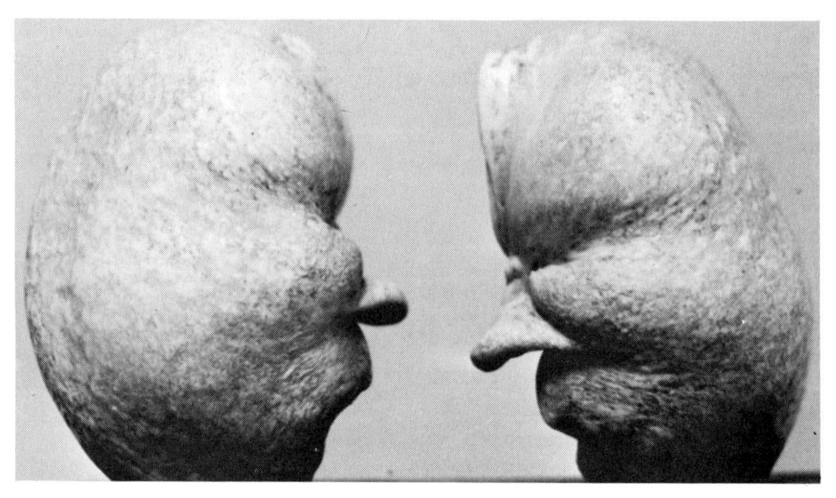

Bottle-nosed dolphins (Jane Burton/Bruce Coleman)

smaller cetaceans normally produce no visible spout when they exhale, only a watery snore.

The cavities around the middle ear have a further buffer zone of a network of blood vessels embedded in fat; these can be engorged, or emptied of blood as required, and act as a pneumatic cushion to absorb the changing pressures of the steep descent and ascent during the dive.

The inner ear is embedded in bone at the base of the skull, and contains the semicircular canals and spiral cochlea typical of all mammals. The canals are relatively small, but effective enough for balance in the whale's environment where, as already described, the body is shaped and the weight distributed much as in a ship's hull designed to maintain equilibrium in water. The cetacean cochlea, however, are notable for their large receptor cells, as in some other mammals (mice and bats) which register ultrasonic sounds above the range of human hearing. The tympanic bone is flask-shaped and no longer articulated with the bones of the skull as in most mammals. It hangs loosely by connecting ligaments and is therefore less sensitive to loud vibrations received by the solid skull and connected bones. (Cetaceans are disturbed and confused if compelled to listen to very loud noises over long periods, just as we are.) Known as the bulla, this bone is exceptionally hard, and is often the only evidence which survives in fossil or semi-fossil form of a once-living cetacean.

Vocalisation is produced by the movement of air in the trachea and nasal passages. Some sounds are emitted through the blow-hole(s) which are closed except during the brief exhalation-inhalation at the surface. The nasal plug can be controlled at the will of the cetacean to blow bubbles, and to produce clicking and crying sounds audible to the human ear, for the purpose of communication with other dolphins, as well as with a familiar trainer.

Digestion

The alimentary system of most of the large cetaceans is well known from dissection at the flensing platforms of processing ships and factories. The stomach is more complicated than that of man (one large bag) or of the cow (two stomachs). It has three compartments: the first is a reservoir for newly ingested food, and in the blue whale is large enough to contain a ton of krill. Food is crushed and softened before it is fed through a narrow opening to the middle compartment where digestive juices and bacteria break it down. Eventually the resulting pulp is absorbed by the third (pyloric) section where the nutriment is filtered into the bloodstream and the waste passed into the small intestine on its way to the rectum.

Like oceanic birds, seals and other warm-blooded sea animals, the

Tail-fluke of southern right whale (Jen & Des Bartlett/Bruce Coleman)
Bottle-nosed dolphins showing blow-holes (Heather Angel)

Two humpback whales feeding on plankton, their throats (grooved, furthest from the camera) swelling; the back of the heads and upper jaws face the photographer (Al Giddings, Survival Anglia)

cetacean has no access to fresh water, and how it copes with its water balance is unknown. The fresh water in its body can only derive from its food.

It will be appropriate to mention here the internal parasites which invade the cetacean viscera from its food; some penetrate far into and damage the tissues and neural complex. The oceans are a fertile zone for the propagation of numerous nematodes, trematodes and other small parasites which, in adult or intermediate form, or as ova, are swallowed with the food and sea water, and usually within the body of the live fish or krill. Some are harmless, even beneficial as nourishment along with the fish-host in which they were living. But the bodies of almost all cetaceans examined are found to contain variable (often considerable) numbers of tape and other stomach worms, also flukes, larval cysts in the blubber and intestines, hookworms and lung mites, which are harmful and have a debilitating effect.

Certain nematode worms burrow into the heart, kidneys, lungs and air passages where they may damage vital neural connections and can cause terrible pain. Afflicted individuals will bash their heads against hard objects under this stress, and as a last resort throw themselves ashore to die. On the other hand, some cetaceans, carrying enormous masses of internal parasites, seem to remain fat and healthy. Ommanney

(1971) describes how Zulu workers on the flensing platform revelled in the blood and mess when a whale was cut up, and 'took the long chains of tapeworms and decorated themselves with loops of them over their shoulders and chests, over their arms and like girdles round their waists, and thus attired did their war dance, whirling their arms aloft, twirling their knives'.

Reproduction

From the many thousands of whales slaughtered and cut up a fairly complete picture of the reproductive cycle of the large cetaceans has emerged. The baleen female is believed to become oestrous for a short period once a year, usually in winter while in the warmer waters of her ocean range and still in good condition from feeding all summer in higher latitudes. The ovum, even in the giant blue whale, is only a fraction of a millimetre in diameter at fertilisation. During mating the pair swim with flippers touching, belly to belly; as a rule the male on his back beneath the female, but sometimes the pair lie sideways.

Swimming among and filming southern right whales off the coast of Patagonia, William Curtsinger watched their act of love: 'Male and female stroke each other's side with their fins. Shifting his great form, belly up, the male's penis emerges, grey-pink in the half-light, sways cobra-like, trance like, and enters her. Belly to belly they come together, moving gracefully, fluidly, forward. Their great bodies shake; then, finished, the male turns upright, moves his flukes over the contours of her back. They swim out of sight, side by side.'

The gestation period varies with the species, but in general is about one year, rather less for the small porpoises. One young is born (twins are rare). Four pairs of rudimentary nipples have been found in some porpoise embryos, indicating that their remote ancestors evidently gave birth to quite a large 'litter', probably as land animals.

The calf usually emerges tail first and as soon as it leaves the maternal body its mother whips round, thus severing the short umbilical cord, and noses it to the surface to take its first breath of air. The twin mammary glands lie one each side of the genital opening and the cow lies at first on her side so that the calf can both suckle and raise its blow-hole above the surface to breathe. As it becomes a good swimmer in a short time, it soon takes the teat from below, with its mother in the normal upright floating position. In all observed 'suckling' the teat shoots forth as soon as the calf touches the slit in which it lies. Many cetaceans have the lower lip projecting awkwardly beyond the upper one, making sucking by the calf difficult; the calf does not suck however. It grasps the protruded teat between the tip of its tongue and the upper lip and thus making a watertight junction receives the milk as it is squirted into its mouth by muscular action of the cow—evidently with some force, for some may trickle out of the corner of its mouth during the process. Because the calf must breathe every ten to twenty seconds, teat-feeding is by brief periods of a few seconds; having taken the contents of each breast in turn, the calf goes to the surface, and the emptied milk-containers are retracted within the warm maternal body. Here they are recharged with the abundant and rich cetacean milk which contains about 40 per cent fat (the richest domestic cow's milk contains 5 per cent fat).

Breast-feeding lasts from four to fourteen months, as detailed under species in this book. Maternal love and care lasts much longer, possibly for life in some species. Young calves, orphaned at an early age, may be adopted by other members of the family or pod. It is suspected that the so-called auntie which—as observed in captivity—will act as midwife during the parturition, and will push the new-born to the surface to breathe if the mother delays to do so, may well be an older sister or other close relation.

This loving care has the important function of guarding the calf against dangers at sea. In the aquarium tank individuals of a different genera will sometimes act as midwife or nursemaid. These services extend to helping a mother to expel a dead child or an afterbirth protruding from the vagina, by gently and persistently tugging it.

The genital-mammary gland region in the female is an area of considerable sensitivity, controlled by numerous muscles, some of which give mobility to vaginate and invaginate the teats. The vaginal opening in dolphins is freely flexed to grasp and carry small objects such as a ball or ring; this may be partly masturbatory behaviour—normally such objects are moved about in the mouth, or pushed with snout or flipper.

Captive dolphins frequently masturbate, no doubt more often than when living wild and free; they must be very bored much of the time in their tanks. Females rub the vaginal orifice against any suitable tapered object, especially the back-fin, flipper or tail of a companion, and may even grasp it with the vaginal muscles. Males similarly rub the region of the penial sheath against an object, and will insert the penis into crevices and crannies, as for instance under the carapace of a swimming turtle. They can be stimulated by a trainer's hand-signals to erect the penis—as an experiment. It is at times embarrassing in the eyes of the public when erection occurs quite naturally at a seaquarium. 'Randy old man!' was one comment I heard at an aquarium in Canada; my reply was 'Not enough for the poor prisoners to do. How would you behave, if well fed but shut up day and night with nothing to do except play with attractive members of the opposite sex naked all round you?'

The cetacean penis is erected by muscular action rather than engorgement with blood as in man, and is extended from its sheath suddenly and rapidly, at the will of the male. It is nearly 9ft long in the mature blue whale. Intromission is usually brief, but frequently repeated during the period of acceptance by the female. The retracted penis disappears completely within the warm body, the opening covered by a muscular flap, and the belly resumes its smooth streamlined shape. The testicles are entirely hidden at all times.

Body colour and markings

In some cetaceans, as in other animals, the colours of the body often become more vivid during the mating season, as for example in the common and other boldly coloured and marked dolphins. Evidently these handsome dolphins are able to perceive these colours, otherwise they would have no meaning when specially assumed at that time. But other species which are always dull-coloured may see only in dull tones of grey, black and white. River dolphins, living in muddy water, are plain-coloured and likely to be colour-blind (one is quite blind); their inconspicuous colour serves as a camouflage against detection by enemies. They communicate among themselves by sonar.

Does the bold piebald orca, fearing no living thing save man, see only in shades of black and white? The pattern of its white panels would seem to call attention to its presence, and may be useful both as a warning to other sea-creatures as well as a guide to its own kind. The pattern of

Killer whale or orca (Charles Jurasz)

Commerson's dolphin (Francisco Eriza/Bruce Coleman)

body markings must have significance as identification between individuals in a pod and species, to bring together mother and child, male and female at mating time, and in swimming together; and also repelling where there is antagonism between males and perhaps between related species.

Most cetaceans are paler or quite white beneath, but usually this whiteness stops or dwindles towards the genital region, designed as if to direct the attention of the suckling calf to the nipples, and the male to the vaginal opening. Whiteness is said to attract fish, and this may be another reason for the paleness of the belly; also perhaps for the white of the underside of the flippers found in many cetaceans, notably in the long-finned humpback whale. On the other hand some successful species are quite black all over.

Sonar: Sound navigation and ranging

Although the cetaceans have no visible ears they can hear sounds in water.
Aristotle, 384-322 BC

Most cetaceans appear to have two voices or vocalisations. The more sonic (audible to human hearing) consists of a vocabulary of clicks and whistles uttered both above and under water, including bubble-blowing, as already described. These vocal signals are supplemented with mechanical ones made by jaw-snapping, slapping of flipper or fluke, and crash-dives after leaping at the surface.

A dolphin experimentally 'blinded' by placing suction cups over its eyes finds its way about the aquarium without bumping into the walls or any other obstruction—even fine wires—placed at random in its path. It is guided in darkness towards the object it is seeking by the echoes it receives from its own clicks and other emitted sounds. These echoes also inform it of the nature of the object they bounce back from; for example whether it is large, small, hard, metal, inorganic or living, or the wide floor of the sea; each reflects back a distinct echo which the cetacean brain analyses with split-second speed and accuracy, and acts upon accordingly.

As cetaceans have little or no sense of smell, their highly developed sonar must replace olfactory recognition of their food or other living things when they approach these deep in the twilight or complete darkness of abyssal ocean. The great sperm whale uses sonar to locate the giant squid and other benthic fish as it quarters the sea floor more than 500 metres deep, where the only light it may see with its small eyes may be the luminescent flash emitted by certain deep-living fishes and crustaceans. This is quite a feat of fine interpretation of the echoes of its vocally produced sounds, since, if the fish rests quietly on the bottom, the whale must distinguish it from the ambient echo it receives from the ocean floor. Once the cuttlefish or other animate object moves, however, the whale can pursue and home upon it by sonar more certainly; and it may then be guided by detecting vibrations and water-movement set up by the escaping prey.

Some dolphins living permanently in muddy rivers have almost rudimentary eyes without focussing lens. But seals inhabiting murky sand-obscured shallow seas, and giant whales feeding in polar seas opaque with plankton, also find their food without much help from their good seeing eyes. Seals in both the Arctic and Antarctic, diving deep under thick ice, emit squeals and other sounds, and in near darkness locate their food by interpreting the echoes returned to the internal ear. It is possible, however, that seals, which have a strong sense of smell—by which in air they identify their child and other seals by the distinct individual odour—can detect prey under water by olfactory clues. Seals, like walruses, have cat-like whiskers or mustachios (vibrissae) attached to the mobile lips, and use these tactilely in the underwater darkness, and in beds of thick seaweed, to locate food, including (in the case of the walrus) clams and other shellfish which normally do not move about, although their antennae set up miniature currents as they wave them to draw their planktonic food towards them.

Hair has virtually disappeared from the cetacean body; but significantly the upper surface of the beak of the near-blind bouto dolphin living in the turbid waters of the Amazon carries a good scattering of short bristles which evidently have a tactile function. So too may the very few hairs which survive here and there on the upper surface of baleen whales (the humpback usually has a single hair crowning each wart or bump close to the upper jaw—a relic of the snout hair of the cetacean foetus which is normally moulted before birth).

The fascinating story of the discovery of sonar in animals had its beginnings in simple observations on the voices of bats as long ago as the eighteenth century. In the present century it has been studied extensively. More recently it has been discovered that oil-birds and swiftlets deep within caves and limestone 'chimneys' echo-locate their individual nests attached to the rock walls swiftly and accurately in total darkness.

The use of sonar under water by cetaceans was accidentally noticed early in World War II, when Arthur McBride, in charge of the world's first dolphinarium in Florida, was trying to catch bottle-nosed dolphins. He discovered that when a school was driven towards small-meshed nets extended across a narrow arm of the sea, the dolphins emitted excited clicks as they dashed at high speed towards the nets, but could not be forced into them, and each time turned back within about 30m (100ft) of the obstacle. However, when he used a net with a mesh wide enough to admit a dolphin's head he had more success. We know know that the rapid clicking sounds emitted by each dolphin were reflected back to the receiving apparatus of its auditory system by the closer mesh—and probably by the minute air-bubbles trapped in the cordage. But clicks passed through the larger mesh without echoing back and deceived some—but not all—of the agitated dolphins into believing the way was clear to slip through.

This discovery of echo-location in dolphins during a world war stimulated the development of underwater listening and sound-recording devices: echo-meters (depth-finders) and hydrophones for the detection of submarines, anti-submarine nets, mines, wrecks and other submerged objects. Later asdic (directional sonar), long used by cetaceans, was invented.

It was soon realised that not only cetaceans but many species of fish and crustaceans produced their contribution to a considerable and often confusing chorus of sounds picked up by the listening and recording hydrophones. It became necessary to devise means of identifying, separating and recording only the sounds and echoes from man-made objects hidden beneath the surface.

During the spawning season grouper fish groan in a noisy chorus, and snapper shrimps crackle like fireworks. Above these extraneous noises, the searching cetacean emits both sonic and ultrasonic clicks under water to assist in navigation and in locating food. Generally, the hearing limit of fishes is below the ultrasonic pulses of cetaceans, so that the approach of a school of whales or dolphins exchanging information between them does not immediately scare away the fish.

Air, trapped in the breathing apparatus, is recycled into and out of special air sacs lying between the nasal plug and the reservoir of the lungs. In this to and fro movement the muscular flap or lip of the air sac above the melon opens out to accommodate the air flowing from the lungs. The soft pneumatic melon swells outwards, but on compression again the air vibrates this lip against the bony wall of the trachea as it returns to the lungs, causing those clicks which are bounced or reflected by the bones of the skull and refracted outwards through the melon.

Something of the sound produced by this internal movement of air between lungs and nose and ears can be simulated by pinching your nostrils shut and closing your mouth tightly, then trying to breathe: a little of the air still present in your lungs will, under pressure upon the eardrum through the eustachian tube, cause a faint click heard within your head at each attempted inspiration or exhalation.

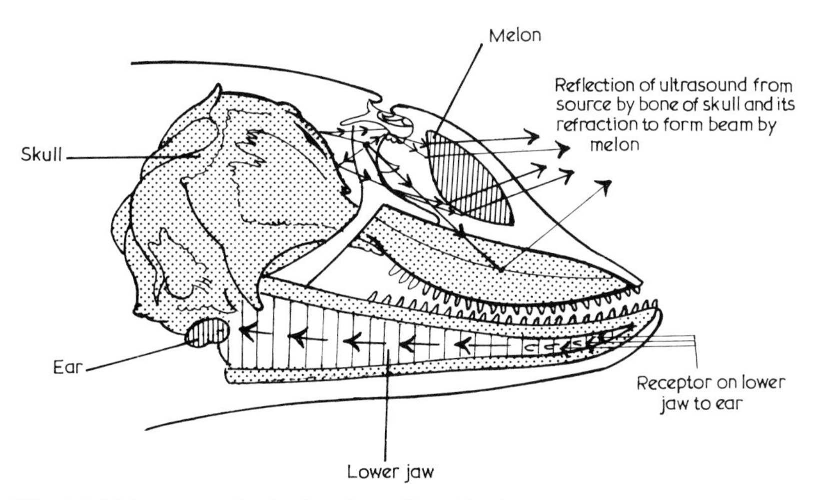

The dolphin's system of echo-location—Norris's theory

The cetacean does it more efficiently, using the same air over and over again while submerged, and by varying the amount passing through the vibrating lip is able to control the pitch or wavelength. Dolphin clicks have been recorded on sonagrams from a low of 20 clicks per second to an ultrasonic high of 800 clicks per second—far beyond the range of the most acute human hearing.

Tests with listening devices held against every part of a dolphin's body have proved that under water the sound is directed from the voice-box of the air-sacs and lungs through the distensible melon and straight ahead of the dolphin's snout. The emitted clicks are highly directional, beamed in a narrow arc in front of the dolphin (about 10 degrees each side of the beak); in order to scan a wider angle the dolphin sways its head from side to side when echo-locating a distant object.

Few clicks per second tend to produce low notes over longer distances —these are the searching sound beams of low frequency and long wave length; the returning echo gives the dolphin, so to speak, a wide and far but somewhat imprecise acoustic view of the scenery. To obtain a more detailed sound picture of an object in the foreground of this waterscape, the dolphin homes in upon the object by increasing the frequency, using a higher note and shorter wavelength. At last the dolphin, like the flying bat in the night, bombards its target with high notes giving it the exact position of the interesting object, be it a companion or a moving fish it is about to seize.

It now seems likely that dolphins and probably most cetaceans during their daily travels use several frequencies almost simultaneously, sound-scanning their underwater world and its contents rather as we view both the landscape and objects in it with seeing, discriminating eyes.

Loud, low frequency (sonic) calls are received through the pinhole channel leading to the inner ear, but high frequency ultrasonic pulses and echoes reach the inner ear by other channels—through the outer wall of the melon and through small nerve-holes each side near the tip of the lower jaw. Here a minute canal filled with oil conducts sound through the bone to each ear-drum.

The enormous reservoir of the sperm whale's gigantic head, containing the prized spermaceti oil, has evolved as a specialised sound path for echo-locating its benthic food at great depths. In effect it is a highly concentrated directional sonar beam machine, like the modern but more clumsy asdic used in ships today for echo-locating objects under water, including shoals of fish and the whales themselves. The cachalot's asdic is an auditory searchlight which can be flashed through the darkness directly ahead of the whale. As it searches the deeps, and when communicating with other moving objects, such as other dolphins, sharks and ships's hulls, the beam is sonic—it can be heard on a ship's hydrophone. The cachalot's eyes are placed too far apart under that massive headpiece for it to see directly ahead of its snout; but the huge melon almost certainly is a sophisticated apparatus for concentrating its own ultrasonic

Cachalot or sperm whale—artist unknown—probably from a stranded specimen

sounds and receiving back the echoes as it closes upon its prey in the darkness. Only thus can we explain the success of this whale in obtaining its food in a gloom more profound than that of the black midnight in which the near-blind bats hunt.

The cachalot's directional sonar also enables it to measure its distance from the surface or the bottom of the sea as it rises or descends; and probably its sound beam, emitted with the body at an angle to the horizontal, is bounced back and forth in a zigzag for a very long distance ahead, perhaps for tens of miles.

By contrast the baleen whales, which feed at or close to the surface, have little or no melon, although they have at least as great, or a greater, repertoire of underwater sounds, signals and songs.

Modern development of clinical apparatus for sound-ranging to produce echoes useful to man include ultrasonic exploration of the internal body structure, which may be more effective and less dangerous to the human body than X-ray examination. This apparatus directs into the subject a short pulse of ultrasonic sound of between 100 kilocycles and 15 megacycles per second. Somewhat as seismologists explore the geological structure of the earth's crust, by recording the echoes bounced back from deep strata layers by explosive sounds applied at the surface, the successive waves of echoes from the different densities of the body parts are recorded on an oscilloscope, and can be charted visually as a cross-section of the investigated area.

In this way an ultrasonic image of the subject of study can be produced, whether it is a seismological picture in depth of the geology, or a cross-section of the detail of that portion (or the whole) of the body being examined ultrasonically. But this sophisticated method of sound-scanning to reveal the structure of a living body, by interpreting the echoes from bone and flesh, from heart, lungs, liver, etc, is only successful at present if the surface of the object is placed in direct contact with the sound transducing apparatus—to eliminate natural dissipation of sound in air, or—and this is the point of the comparison—both the apparatus and the subject are placed close together under the water with the transducer pointed at the subject.

Beluga or white whale underwater showing the melon (Pat Morris/Ardea Photographics)

Beluga or white whale underwater, demonstrating its very different appearance from above (Pat Morris/Ardea Photographics)

But the cetacean, thousands (probably millions) of years ago, had achieved this in-depth analysis of the echoes it receives from the ultrasonic sounds it directs at objects under the water. When investigating in this way another dolphin, fish, seal, or live creature, it is almost certain that it 'sees' acoustically the shape, size and internal structure of the animal, even as it judges accurately its exact distance. It probably hears its heartbeat.

Just as most of us are normally able to recognise the small noises and movements of a familiar person in the house without seeing them, it is probable that the cetacean recognises the acoustic picture it receives from a familiar person, as a blinded dolphin does a companion or its trainer swimming within hearing range of the echoes of its probing sonar.

'I hear you, but do I know you?' the cetacean sound-scanner asks through the opaque water. 'You hear my signals and know me by them. I am calling to you, my friend. Answer, so that I may recognise your familiar voice.'

All these controlled signals and echoes—sound messages—are emitted and received and interpreted in pulses lasting split seconds, milliseconds in fact! And some believe that even the mood of the animal —hostile, happy or normal—can be gauged by ultrasonic scanning.

As in echo-location in bats in air, the dolphin under water has to distinguish between emission and echo. Theoretically there is a time interval (silence) between each emission and its return as an echo. To the human ear the sonic click of a dolphin, heard in air or through a hydrophone below the surface, lasts usually less than a second. When a recording is played back slowly and appears visually (analysed by oscillogram) each sonic click proves to be composed of many separate clicks per second, while ultrasonic clicks consist of 800 and more sound pulses per second. But can the cetacean brain conduct the same deep analysis as the oscillogram, separating output from input, emission from echo?

As yet such millisecond analysis by the cetacean has been impossible to prove. But if the echo of one millisecond click coincides with the emission of the next pulse, presumably the merged or overlapping series will create a different (continuous) sound which is more effective for navigation than a single click and its separate echo.

To puzzle our understanding further, the cetacean is able to utter

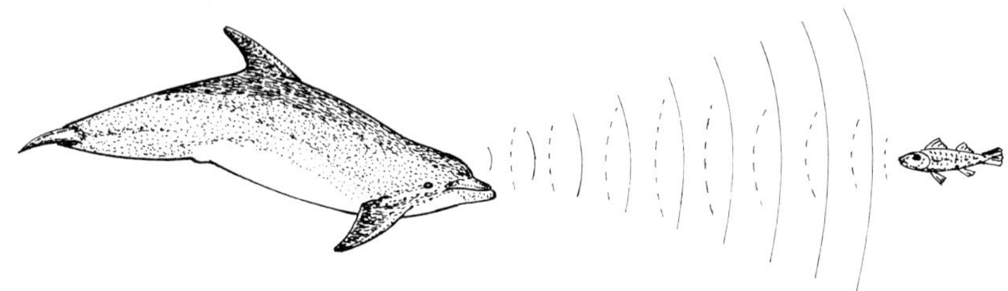

Dolphin echo-locating fish

controlled clicks and whistles simultaneously, both in the air and under the water.

In captivity a dolphin will release through the blow-hole under water a single large bubble causing a loud plop. This is regarded by some observers as a query: in effect—'Here I am. I am listening, where are you?' A small stream of bubbles is probably a different signal, but in both instances the cetacean has the air under control, and is deliberately signalling by bubble noises. Humpback whales do this (page 149).

Cetacean vocalisation is very varied, each species having its special sonic and ultrasonic and 'song' channels which it listens to, mostly ignoring those of a different species. But dolphins are great mimics, and in captivity will relieve their boredom by practising new sounds they have picked up from other animals, including man.

Common dolphin showing blow-hole and teeth (Heather Angel)

Two dolphins visually isolated in separate tanks used both clicks and whistles to become acquainted. They held a cetacean conversation. When one clicked or whistled, the other was at first silent, listening politely and later responding while the other was silent. But sometimes when one clicked the other would whistle at the same time. As if amused by this exchange the clicking dolphin might next mimic the whistle pattern of the other, and make a duet.

John C. Lilly, studying sonic and ultrasonic emissions in dolphins in his well known attempt to discover if the dolphin has a language of its own, has recorded on tape a considerable vocabulary of delphinese 'words'. He considered that in general the whistle is usually confined to occasions of varying degrees of distress or agitation in the particular individual. The response of nearby dolphins is to swim at once to the aid of the whistling one, and thrust it to the surface to breathe, even if it is not obviously needing air. The first breath taken by the agitated dolphin is usually followed by quite a conversation between the group, an excited whistling and clicking which may go on for some time and, thinks Lilly, 'is evidence of meaningful exchanges in the vocal sphere'.

When I met Lilly in 1963 he told me he believed that one day someone would break through the language barrier between the talkers in the sea and on the land. He was still baffled by the swiftness of the dolphin's response to human conversation, and their skill at mimicking human speech, but as they have no vocal chords, their imitation was squeaky and strange and delphinese. He concluded: 'It's time I went away and studied my own mind, for the wonder of the dolphins poses the question of where we humans are heading? Apparently into darkness, with the wiping out of an underwater civilisation that has managed the oceans with superb poise and conservation for millions of years before upstart man appeared and began to ravage earth and sea!'

Sperm-whale-hunters today locate their prey with asdic, beaming sound deep to where the cachalot feeds on squid. When this whale hears the ping of the echo-sounder it sometimes gives a click to match each mechanical ping, as if answering, and curious about, this strange call from the keel of a ship high above. The cachalot can be detected at some distance by the ship's sonar as it clicks at regular intervals of two emissions per second, which is a slow rate and therefore long ranging and suitable for seeing acoustically for several miles, as when the whale is calling up its companions, or locating food.

The sensitive hearing of cetaceans makes them particularly vulnerable to continuous loud noise which smothers their sonar signals. Hunters make use of this fact to drive schools of the sociable dolphins and pilot whales ashore, by beating the surface of the water, revving engines, banging tin cans, etc. The confused animals crowd together and pile up in shallow water. Young whales have been isolated from a pod at sea by circling them in fast rubber dinghies powered with screeching outboard motors. Unable to hear the calls of its companions beyond the circle,

Stranded pilot whales in the Orkney Islands (Topham)

the individual remains in the centre as if dazed, panting at the surface and seldom making an effort to escape by diving.

Wood-wind music, gently played, attracts dolphins to swim around a sail boat stationary or slowly moving at sea, where the waves are quiet and there is no engine to suppress the notes. They seem to listen with enjoyment and certainly are very curious at first, poking their heads out of the water to stare at the source.

The fact that many cetaceans breach and breathe at the surface in unison, with split-second synchronisation, whether in pairs, pods or schools, has special significance in communication. This synchronisation enables them to hear and talk together, particularly in the long intervals under water, approximately ninety-eight per cent of their time and lives. The noise of breathing out of time with each other would confuse and interrupt these conversations, and reduce their chances of hearing the approach of an enemy. Paul Spong believes that orca adults train their children to breathe in respiratory units; that is, 'All together now children'.

Humpback whales, cruising alone at some distance from each other, seem to make a louder blast, as if signalling their presence over this longer distance. But when swimming in a pod close together the blast seems quieter as they breathe in unison, then submerge and resume their companionship below the surface; and the observer has the feeling that the simultaneous blast is simply an involuntary punctuation in the underwater talk and gossip and other intimacies of a pair or family group. Dolphins, pilot, false killer and sperm whales do it. It seems to be cetacean etiquette not to interrupt the submarine conversation and deployment of sonar by breathing, except at agreed timed intervals.

The whale's way: The ecosystem

Where great whales come sailing by,
Sail and sail, with unshut eye,
Round the world for ever and aye.
Matthew Arnold

To survive, reproduce and evolve it is necessary to eat. We may skip the passage of millions of years of this basic driving force which compelled the gradual return of the whale, as an air-breathing terrestrial mammal, to the bounty of the sea, as the pressure of competition with other animals for the food available on land grew more intense, perhaps over ages of prolonged drought or cold. Nor do we know in what form the ancestors of our present whales first returned to the water. The fossil evidence leaves a wide gap. Probably like the otter today it waded and swam and dived in river, lake and the sea's edge in its early efforts to catch fish and other edible water creatures. Becoming more amphibious and stream-lined, it began to trail its hind legs, like the seal does today, and to develop these and the tail into a single effective swimming unit, like the dugong and manatee, using its forelegs only to drag itself ashore to rest, and give birth to its young. Eventually, although still breathing air and giving birth in shallow water, like the gray, humpback and other whales, it gave up the land altogether.

Adapting, evolving, changing its form, size and habits it spread far over the oceans, themselves changing with the shifting of the earth's plates which carry the world's continents. And during the several glacial periods it was far easier for the creatures of the sea to swim away

Gangetic dolphin—origin of artist unknown. This dolphin has no eye and is quite blind

from the advancing ice than it was for land-bound animals, many of which were overwhelmed by it.

Some whales were left behind, or returned, to inhabit rivers and estuaries—the freshwater dolphins we know today. Others dispersed into remote corners of the warm oceans and became sedentary, certain species of dolphins which are rarely seen, and, according to Bruyns, several have never yet been examined in the flesh, although he has seen and described them at sea.

Most remarkable of all has been the adaptation of the cetaceans to become so large as to exploit and withstand the cold waters upwelling at the edge of glaciers and frozen oceans. Strange as it may seem this turbulent near-freezing water is richer in food than any tropical sea; so rich indeed that the water is turbid with layers of life, making it difficult for underwater photography. To make full use of the vast plankton soup of the cold sea, whales, as we have seen, have developed the special straining devices of baleen plates, to separate the living particles from the water, and have grown in size to become the largest mammals the world has ever known. For largeness and bulk are most economical in conserving energy because of the lower ratio of body surface to cubic capacity conserving internal heat. The tiny body of the humming bird needs to be replenished with its own weight in food every

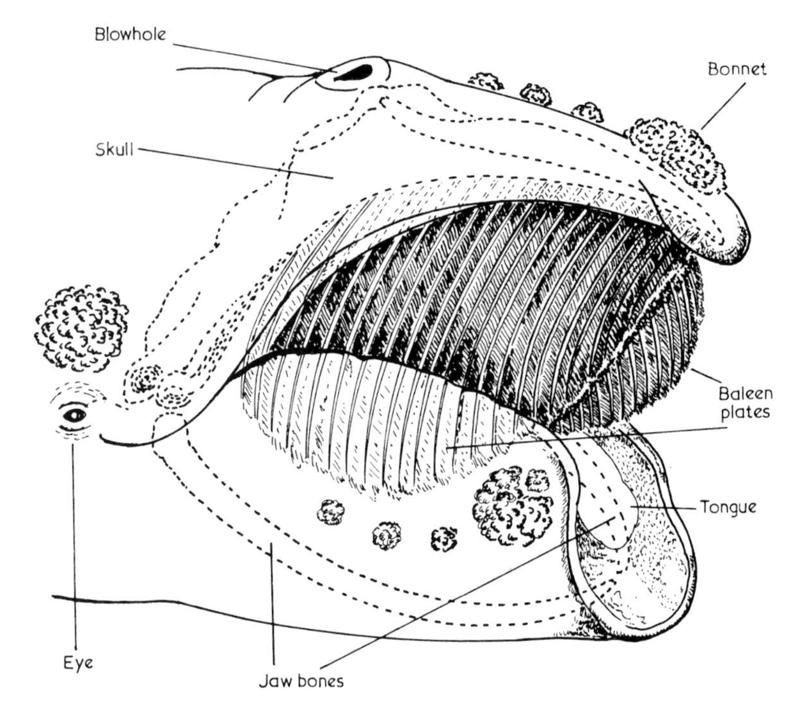

Baleen plates and jaws of right whale

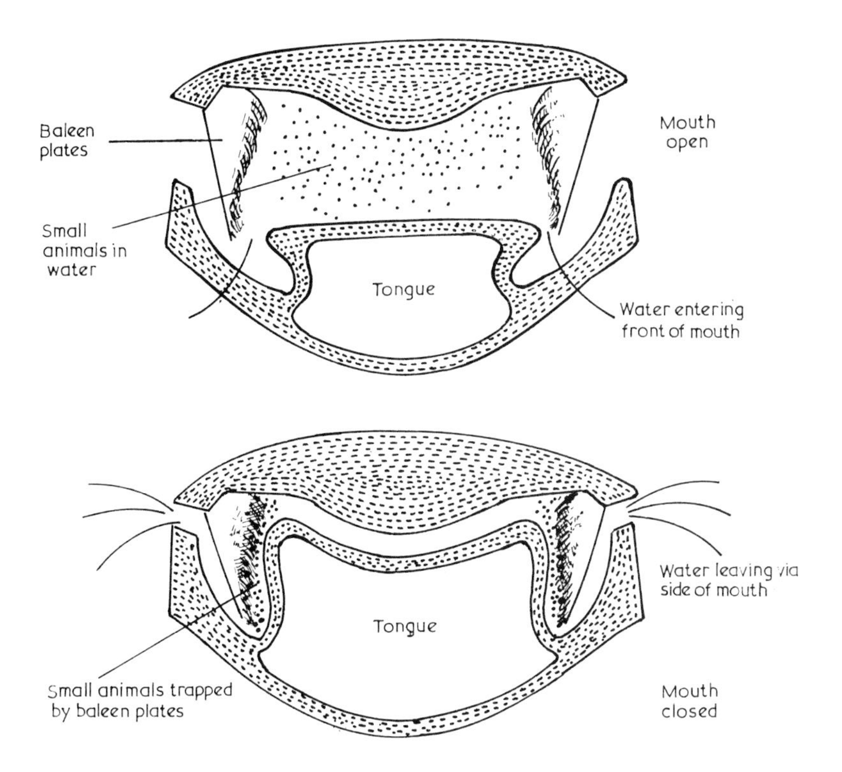

A rorqual's mouth showing how baleen whales feed

forty-eight hours in order to compensate for the loss of energy (heat) through its relatively enormous body surface, whereas the bulky ostrich can fast for many days and remain in good condition. So too, the large whales do not lose body heat easily: they are known to fast for many weeks, if not months; but during their intensive feeding in polar seas a few degrees above freezing point they are able to put on a thick layer of fat to last them for the rest of the year.

The Ocean Food Chain

The long sunlight of the polar summer day encourages photosynthesis. Nutrients and salts, upswirling from the cool depths to the warmer surface water, feed and energise the minute phytoplankton (diatoms, plants and algae) which multiply by rapid division, providing in turn basic food for the zooplankton, popularly referred to as krill, tiny animals in perfect and larval form, such as shrimps, crabs and other crustaceans, copepods, jellyfish, sea worms and other free-swimming creatures. Yet the water temperature in these regions, even in summer, remains only a few degrees above freezing, slowing down the growth and maturity of these minute animals so that many species live comparatively longer than their tropical relatives; some are able to survive in their larval or

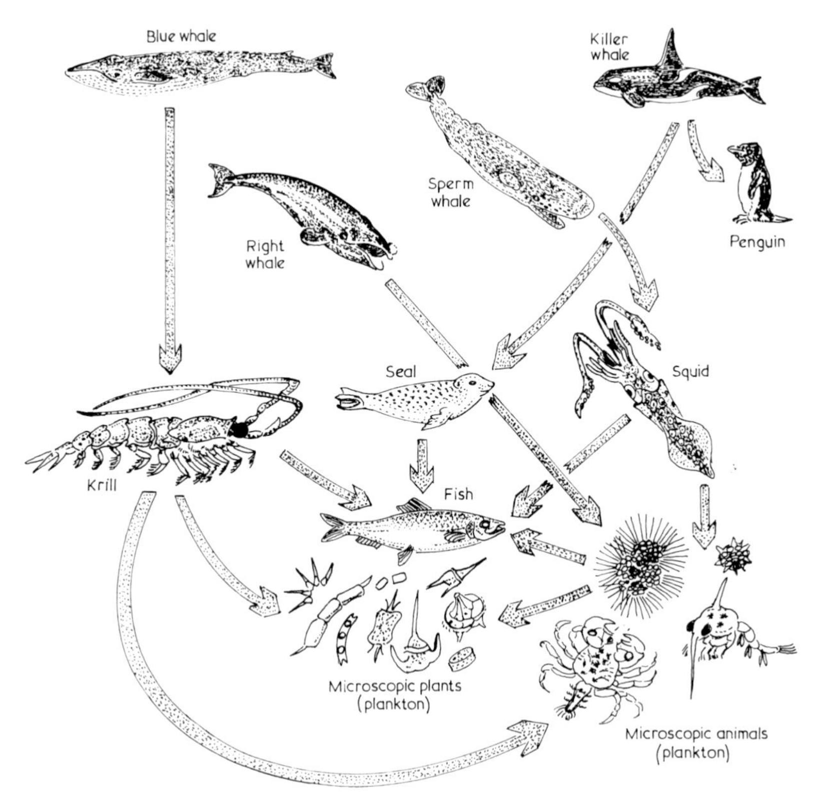

Cetacean food chain

immature state at near-zero temperatures under the advancing ice of the long polar winter. At that season some plankton species are in effect cold-stored alive for the return in spring of the sun and the great whales which, with the retreat of the ice, will feed and fatten on this swarming manna of the sea.

The food chain of ocean is a vast web based on this abundant phyto- and zooplankton (from the Greek words meaning 'animals which drift') drifting in masses or loosely spread, now rising to the surface in dense swarms, sensitive to sunlight and slight temperature changes, and carried by vertical and horizontal water movements caused by a combination of wind above and circulating ocean currents below.

Each baleen whale devours several tonnes of this krill every day it feeds in the polar and sub-polar oceans. Sea-birds of many species, hovering, swimming or diving; fishes, including squid and plankton-eating sharks fitted with other forms of mouth sieves (gill-rakers): all these depend on zooplankton for a living. The smaller fish and squid are in turn devoured by larger fish, which are eaten by the seals, dolphins and porpoises. And so on through the food chain or cycle, larger eats smaller, to the sea leopards and sharks, which eat fish and penguins, to the sperm whale which kills giant squid and octopus in deep water, to

The stomach of a whale bulging with krill (Topham)

the killer whale which fears no other creatures, but will feed alike on fish, the flesh of penguin, seal, dolphin and the baleen whale itself.

Constantly from this moving web of sea life there falls to the ocean floor a steady rain of organic matter in the form of faeces and the bodies or remains of dead animals. Some of this waste is devoured, further reduced and excreted by crustacean scavangers of the sea bed; all finally decomposes into elements of phosphorus, potash and nitrogen which, mixed with other mineral salts from the land and air, are carried by the upwelling currents to nourish the phytoplankton drifting in the euphotic zone at or near the sunlit surface. The chain is complete, from plankton to cetacean, seal, fish and sea-bird, and back to the crucible of the sea floor. Each animal has its place in the marine food cycle, the recycling ecosystem.

Approximately 90 per cent of the zooplankton of the antarctic and subantarctic sea is composed of the small red prawn *Euphasia superba*. Some sixteen species of euphasiid prawns or shrimps, copepods (twenty-six species), amphipods and other planktonic animals are found in warmer waters of the southern hemisphere, especially where there is upwelling of cold currents, and attract southern baleen whales to feed during their migrations.

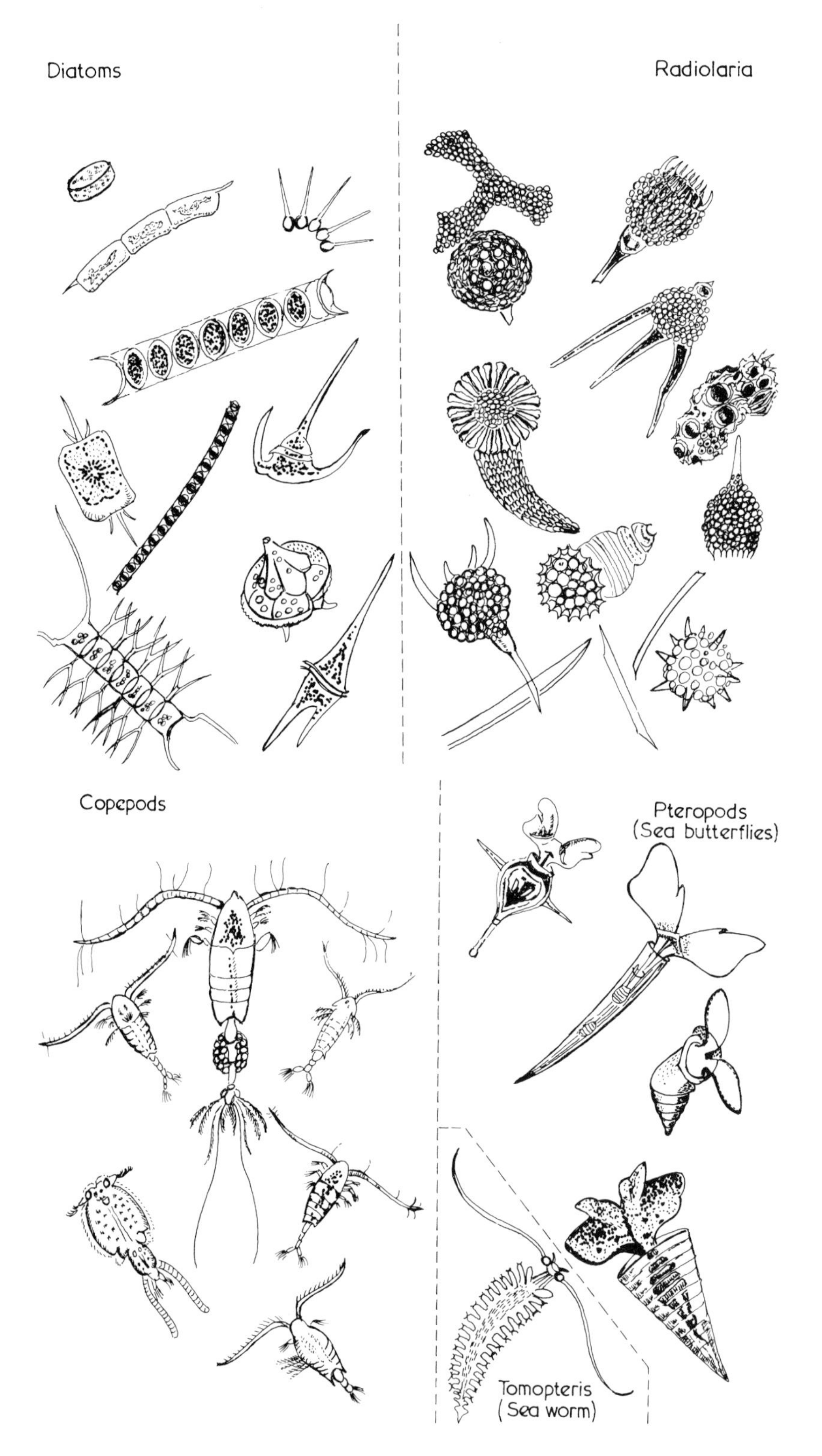
Diatoms
Radiolaria
Copepods
Pteropods
(Sea butterflies)
Tomopteris
(Sea worm)

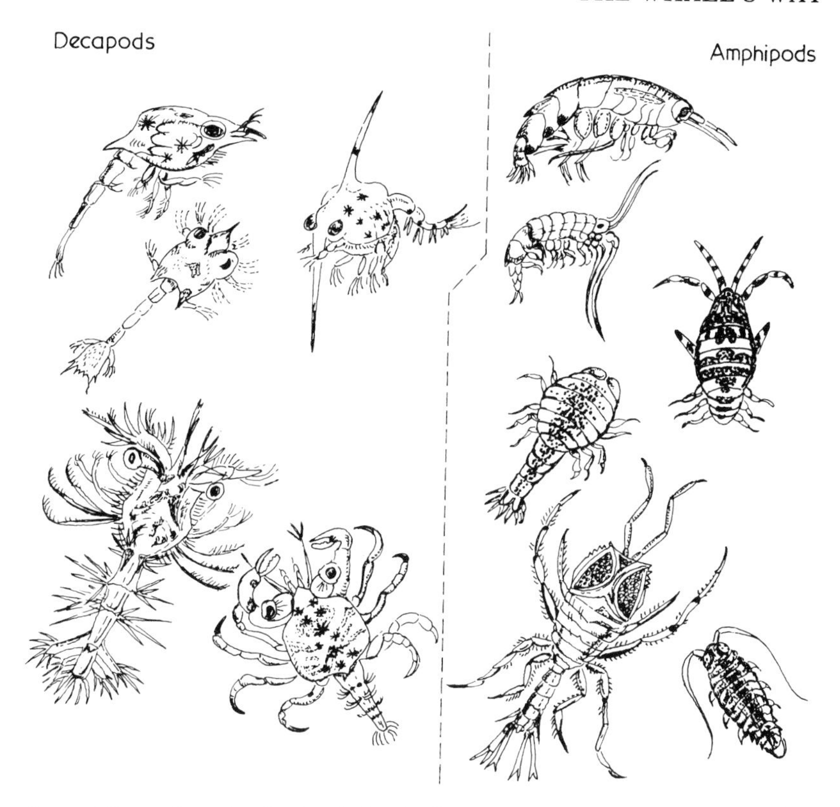

Plants (diatoms) and animals of the plankton

In the northern hemisphere there are few euphasiid prawns in the arctic sea, and baleen whales feed principally on a zooplankton consisting of the pelagic mollusc *Clione*, known as the sea butterfly, present in such numbers that the ocean near the pack ice is pink. With this mollusc are numerous amphipods, copepods, small squid and larval forms of many crustaceans. In addition baleen whales are not exclusively plankton feeders. They and many of the smaller dolphins and porpoises feed freely on the swarming small fry of pelagic fishes, on sardine, smelt, sprat, capelin and other gregarious finger-size species.

It is in search of an adequate food supply that the cetaceans, great and small, make their ocean journeys.

Migration

The cetaceans live in a world that is, or was, largely hidden and incomprehensible to man. Yet their range is three times as spacious as ours, since the oceans occupy about three-quarters of the area of the earth's surface. Through this vast seascape their smooth torpedo-shaped bodies glide with superb ease and grace which we heavy-footed bipeds envy. Like the birds of the air, the cetaceans in the sea move in depths of many metres as well as horizontal distance, sinking and rising at will

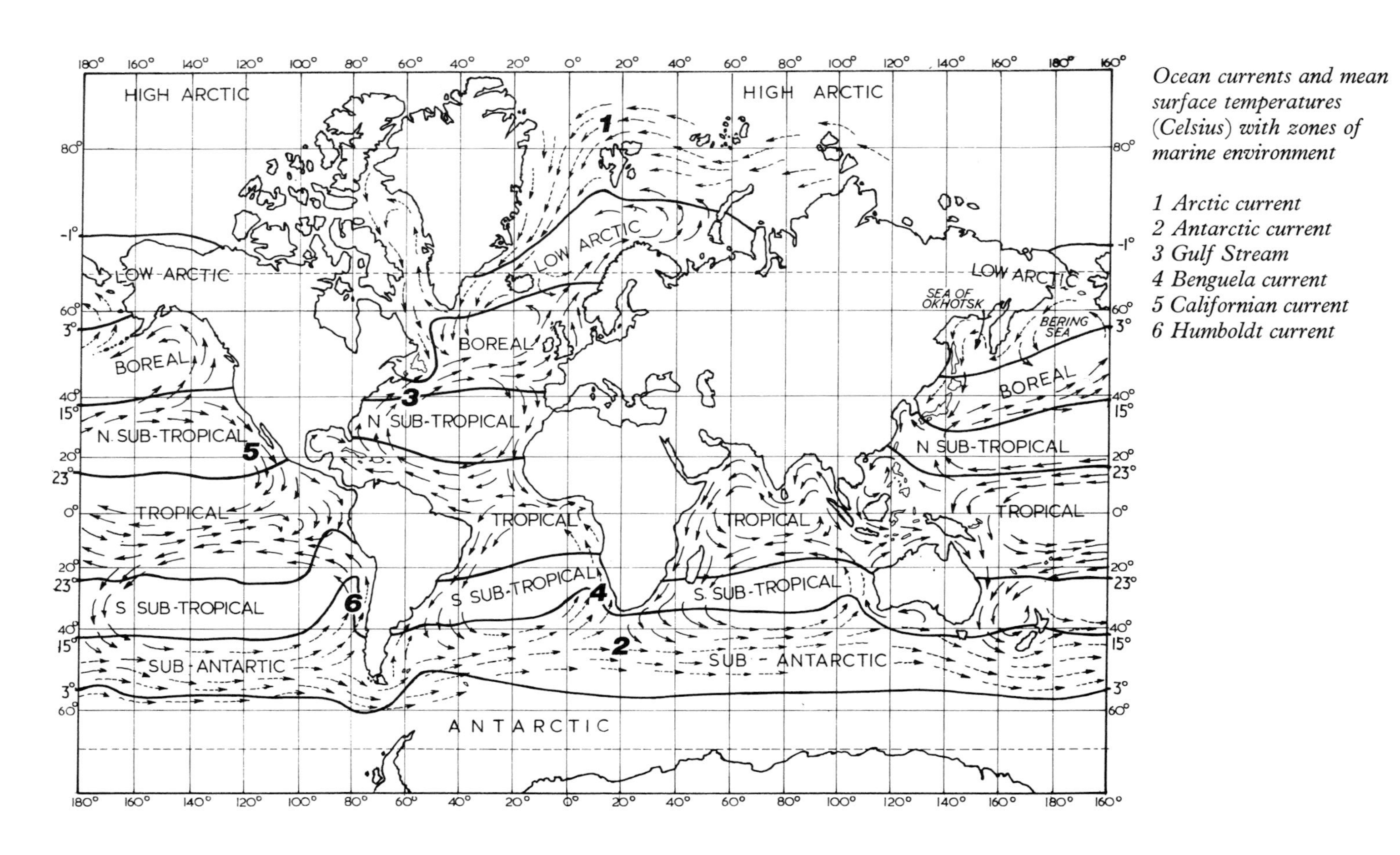

Ocean currents and mean surface temperatures (Celsius) with zones of marine environment

1 Arctic current
2 Antarctic current
3 Gulf Stream
4 Benguela current
5 Californian current
6 Humboldt current

through the well marked zones and layers of the deeper sea, each with its characteristic marine fauna from which whale and dolphin can select their food, and try to avoid their enemies. They glide downwards or upwards at an angle, periodically rising to a surface lit by the sun, moon and stars, to renew the oxygen in their lungs and blood. The sea may be raging, the wind like Masefield's whetted knife, but the cetacean is untroubled by the greatest storms; indeed it is more at home in rough than in calm seas.

Indirectly, however, its life is much influenced by wind. The eastward rotation of the earth, which reaches about 1,600km/h (1,000mph) at the equator, produces the prevailing trade winds, blowing east to west at the equator, between the tropics of Cancer and Capricorn. These winds drag the surface waters and all they contain in a westerly direction. Warmed by its passage through the tropics, the wind-driven water is deflected against the westward continents, turning south-west in the southern hemisphere, and north-west in the northern hemisphere. In the latter, the land masses which almost enclose the Arctic Ocean obstruct and divert the counter trade wind as it swings to the north-east and circles gradually south, cooled now by the flow of cold water from polar ice-fields.

In the southern hemisphere the warm flow of tropical water under the west-going equatorial trade wind produces a genial climate along the eastern shores of Australia, South America and South Africa. But there is open ocean to the south, between latitudes 45° and 60°S. Here the counter current of the west wind drift is driven eastwards unimpeded by land before the almost incessant westerly gales of this zone, appropriately known as the Roaring Forties. The huge mass of water moves fast, chilled by the ice-cold contribution from the Antarctic Ocean, but laden with masses of plankton as well as large icebergs slowly melting as they reach warmer water.

This cold, swift current is split when it strikes the south-western extremities of the three southern continents, before it meets and mingles with the warmer current on the east side of the headlands of Cape Horn, the Cape of Good Hope and Tasmania. The northern portion of this antarctic water is diverted by the mountainous wall of the Chilean coast to sweep northwards towards the equator. Here it is known as the Humboldt Current, a cold, still turbulent flow 100km (60 miles) wide and rich in plankton, plankton-eating fish, whales, dolphins and seabirds—a drifting replica of the food chain of the antarctic. Penguins, normally antarctic birds, frequent this cold Humboldt current; three species of the black and white *Spheniscus* settled on this coast. One, *S. mendiculus*, has reached the Galapagos Islands, the only penguin known to breed close to the equator.

Part of this same cool eastward flowing sub-antarctic current, strengthened and enriched with plankton-laden water from higher latitudes, is similarly diverted north along the south-west coast of South

A large school of pilot whales in Lerwick harbour in the Shetland Islands in 1949. Apart from the odd boat passing near, on this occasion they were permitted to leave unmolested; a whaling station was in operation in the Shetlands in the early 1900s (Topham)

Africa. This is the Benguela Current, where great and lesser whales pause to feed, some perhaps to breed, on migration. Unfortunately even now they are being hunted by one or more pirate whaling ships which, flying the flag of a nation which does not recognise the authority of the International Whaling Commission (IWC) which protects them, slaughter indiscriminately all species of whales and market the products by devious means under that flag.

The south-west point of Australia is another bastion to split the cold east-going current, part of which swings north into the Indian Ocean. Great whales, chiefly sperm and humpbacks, are taken at this point (at the time of writing) on the way to their winter quarters and nursery and mating grounds.

Destinations and Territory

Through the wide oceans and narrow seas of the world the great whales, and many of the smaller dolphins, migrate with a regularity and accuracy man finds hard to understand. It has been easier to study the visible movements of migrating birds, which we know from numerous experiments to be guided by the position of the heavenly bodies, the sun and major stars, as if, like the mariner, the migrant possessed sextant, compass, chart and chronometer. Unerringly, and often exactly to the same day each spring, the long-distance migrant bird arrives to sing again in garden, wood and moorland. But during periods of fog or mist, without sight of the heavenly bodies, the bird will pause on its migration, and wait for the sky to clear; or it may ascend above the clouds to resume its journey under the open heaven.

The oceanic shearwaters on their traditional flyways in both the Atlantic and Pacific migrate in the direction of the trade winds across thousands of miles of sea twice a year between their nesting grounds and their off-season quarters, sensibly crossing the equator from autumn into spring and banishing winter from their lives. In doing so they are following the turbulence of the surface which these winds create, and a correspondingly rich feeding area. On average they cover the distance at the rate of a few hundred miles a day, much faster than the whales which swim along the same route.

Salmon of several species which migrate thousands of miles to feeding grounds in the ocean will return to the rivers and headwaters where they were marked as fingerlings and immatures. It is not known how they do this, but probably by the guidance of celestial signs when far from land. On reaching the coast, they no doubt remember the familiar configuration of the estuary approaches, the estuary itself, and the windings of the home river. It has also been suggested that the individual 'taste' of a river, its distinctive chemical composition and temperature, is also a guiding factor. Do whales find their way in a similar manner?

It is not known precisely where the great whales, feeding in polar, sub-polar and temperate seas, settle to breed and mate in the winter. Humpbacks from the western Pacific coast of Antarctica pass north in fat condition along the coast of New Zealand and disappear through the south-western Pacific islands of Polynesia. En route they are hunted in small boats by the men of Tonga, who seldom take more than half a dozen annually.

Those feeding in polar seas south of the eastern Pacific probably move north with the Humboldt Current as already mentioned—but where they calve is not known. Whales feeding south of the Indian Ocean are probably of the same groups seen at play—evidently courting—or with young calves in tropical waters along the African and Australasian coasts of that ocean, where they were formerly hunted.

Southern right whales assemble with their dependent calves off the coast of Patagonia where they have been studied and their courtship and mating ceremonies filmed.

In the northern hemisphere, humpback whales swim with gray and right whales from winter quarters off California and Mexico, and from

An aerial view of a southern right whale with young (Jen & Des Bartlett/ Bruce Coleman)

southern Japan, north to the Bering Sea and beyond. The gray probably makes the longest migration of all, penetrating in late summer to the ice limits of the Beaufort Sea north of the Canada-Alaska border. Other whales such as the bowhead and beluga make shorter journeys.

It is believed that the humpbacks, which in winter sing and mate and calve in warm shallows between the Hawaiian islands, are the same group which feed in summer on plankton and small fish at the foot of glaciers and snow-filled river estuaries in the narrow sounds of southern Alaska. Here they take up territories within easy calling distance of each other. On the Atlantic side of North America the humpbacks, which sing, mate and calve off Bermuda and the West Indies, almost certainly belong to the school which migrates to the Gulf of St Lawrence and Labrador. Another group (now very few in number) migrates between tropical West Africa and the west coasts of Britain and Norway.

Not many live whales have been individually marked and of these so far only a very small percentage have been reported a significant distance from the site of marking. The general opinion of students of cetacean migration is that the large and medium, if not all species of whales, naturally form discrete groups consisting of family parties and pods (as humans do) which keep to familiar, but restricted, feeding grounds and breeding territories in the latitudes which they frequent, and migrate by an unvarying, accustomed route between these summer and winter quarters.

It is not yet proven that any northern hemisphere breeding group of congenerous family pods ever crosses the equator to mingle with a southern hemisphere group of the same species, or vice versa. Or even that a single individual does so. Generally the two groups live permanently separated several thousand miles apart; when one is breeding in the highest latitude of its range, the other is wintering in the lowest latitude, in the opposite hemisphere. And in this long geographical and ecological isolation most species have developed sufficient differences, usually of size and colour, as to be given subspecific, even specific rank by the taxonomists.

Only in a few species which are found in all oceans of both hemispheres are there no apparent signs of subspeciation—as if these few did cross the equator and interbreed. But this has not been proved. The sperm whale is the best-known example. Others are the grampus, the false killer, pygmy and little killer whales, and the true killer, the orca. But although the last is seen in every ocean, apparently as great a wanderer as the cachalot, there are signs of subspeciation in the amount of white on the brow of the eye (least in the southern-hemisphere orca, which also has a larger melon).

As a general rule the southern race, or subspecies, is larger than the northern, as in blue and fin whales, perhaps because the vast continuous world-circling ocean feeding grounds of the Roaring Forties latitudes may have a richer food chain compared with the smaller areas of the northern feeding grounds which are interrupted by continental and insular lands.

It seems likely that separate pods within the geographical range of

Southern right whale showing barnacles on head (Jen & Des Bartlett/ Bruce Coleman)

the same species, more especially today in the scarcity of the large whales through overhunting, have less chance of meeting and, by interbreeding and the exchange of genes, maintaining the common heredity. This would account for the development of different races within a species occupying the same ocean and hemisphere which have been described by some cetologists.

These finer differences need not concern us here, except to note that we cannot be certain that some immature individuals, not infrequently observed to linger overlong in tropical winter quarters, do not encounter individuals returning towards the equator from the opposite pole. They could meet, at the moment of the year which is springtime for one and autumn for the other. If so, the occasional defection of an individual from one hemisphere to live with a subspecific group in the opposite hemisphere, does not seem to have had any effect on the distinguishing racial differences which have led to the separation adopted by taxonomists.

This study of the evolution of species and subspecies in whales is fascinating, of how groups of one species change in ecological isolation brought about through various influences, including the reduction of the oceans during long glacial epochs. With the shrinking of the sea the cetaceans were driven to live nearer the equator, and some groups were cut off from—and their gene banks closed to—others of the same species on the separation of the oceans by immense sheets of ice. They had become separate species by the time the ice melted thousands of years later and the seas rose to open the gates between the oceans.

In a minor way this evolution can sometimes be visualised in action today during the present phase of the retreat of the polar ice-caps. For example, most of the glaciers in the Glacier Bay National Monument Park, Alaska, are melting rapidly, leaving fiords, closed by ice-walls only a hundred years ago, wide open to the invasion of the sea. As the glaciers continue to retreat several miles inland towards their source under the ice-capped Mount Fairweather Range, the humpback, minke, killer whales, porpoises and thousands of sea-birds have followed the krill and associated fishes to the foot of bergs crashing into the water here; for this drowned valley newly released to the sea is rich in minerals nourishing the plankton.

The whales which have found and utilised this new feeding ground in Glacier Bay are thus enterprising colonists of only a few decades. Their gene for exploration must be strong, and indicative—we may speculate—of some vigour for survival and the development of a race (estimated here to number up to 200 adults) already unique in travelling from the Hawaiian Islands (or wherever they mate and calve) to this geographical extension of their range. But we must leave this speculation for the present, and sum up our present knowledge of the navigational ability of the cetaceans.

Humpback whale in Glacier Bay, Alaska (Al Giddings, Survival Anglia)

Navigation

Does the migrating whale, dolphin or porpoise find its way across the deeps of oceans by glancing from time to time at the sky to read the celestial chart of the moving sun by day, and the stars, planets and moon by night, like the migrant bird already described? If so, it must carry in its brain the equivalent of sextant, compass and chronometer in order to compute its position and navigate the correct course, always aware, consciously or instinctively, that the heavenly bodies which it can see at any given moment in the twenty-four hours move through the sky in a pattern changing every moment. By day it must steer by the sun's position; which argues that it must 'fix' this position by reading its chronometer for time, its sextant for angle to the horizon, and its compass for direction. (By night the moon and planets are surely too irregular and sky-wandering in their appearance; the moon rises up to an hour later every twenty-four hours, and the planets follow their orbits independently of that of Earth.)

Certain vivid stars, however, will greet the unshut eye of the navigating leviathan and its smaller relative, the migrant dolphin. Above the equator Polaris appears bright at due north, and around it other shining stars revolve in a fixed pattern. Below the equator the design of the less vivid Southern Cross is the axis around which the southern-hemisphere constellations move. These patterns were well known guides to the Polynesian people inhabiting the warm coral islands of the central Pacific; they made long voyages of exploration, driven by overpopulation, in search of new lands to colonise. They encountered many species of cetaceans, which they hunted whenever possible, and their adventures with whales and dolphins are part of their traditional tales and songs. The meat was eaten and the teeth were prized as ornaments.

The Polynesians were expert astronomers and knew that at any given minute, hour and day of the year the sun subtended exactly the same angle to the horizon. So too at any given minute, hour and night in the year the star pattern was exactly the same as it had been one year ago to the minute. (The apparent courses of the sun and major stars do change, but so imperceptibly over an epoch of a hundred years that there is plenty of time for the migrant's posterity to adjust its inherited knowledge as generations pass.)

Because of this faithful annual reappearance of the same sun and star pattern and movement in the sky, voyages were usually made at a selected propitious season: the brown-skinned Polynesian, generally a priest trained in astronomy, would carry in his head a chart of the star pattern (perhaps also outlined in paint on the hull of the canoe). With this chart and his knowledge of trade winds, and the signs of land (motion of the swell, currents, sea-bird movements) not yet visible over the horizon, he would steer or direct the navigation accurately over hundreds of miles of empty ocean.

So too, perhaps, the cetaceans, making similar and often longer oceanic journeys, may find their way by all these physical guides, to which we must add yet another. Under the surface the whale is able to examine, visually, or by measuring the distance with its sonar-scanning auditory eye, the topography of the sea floor, with its shallows above lofty marine mountains and elevated plains, and its immense black chasms several miles deep. Like the scenery we remember on our land travels, which guides us in future, the underwater scenery of its migration route will be well-remembered and fixed in the brain of the itinerant intelligent cetacean.

But young whales and dolphins, it may be argued, making the migration for the first time, do not need any of these guides. Like young swans and geese which migrate with their parents and remain together as a family unit for at least one year, the young cetaceans have the advantage of learning, of being shown the traditional route between their place of birth and the limits of the off-season wanderings during the rest of the year.

Although it is interesting to speculate in this way and talk learnedly of circadian rhythm, the biological clock and the genetic computer controlling the animal brain, the navigation of whales and dolphins is as much an unsolved mystery as that of the migrating bird, fish and butterfly.

Co-operation and Competition

We have seen that dolphins are able to co-operate voluntarily with man in gathering fish food. But do species co-operate with other species in the wild, in food-hunting, or on migration? As a rule few do so. It was competition for food and living space which led in the first place to the adaptation and separation of the cetacean family to occupy the different ecological niches under the surface of the oceans and rivers, and thereby diversify into the present seventy-eight species.

Although pods or schools of two or more species will feed in the same area, attracted by the same food, they invariably keep to species groups, each breathing as a rule in unison, diving to feed simultaneously and evidently communicating with others of the same pod with squeaks and whistles (hunting cries). They will hear but not necessarily listen to or take action upon the calls of other species. In the Amazon River the bouto dolphin is seen with the smaller bufeo dolphin, but they play separate roles in the ecology, feeding differently, one in shallow, the other in deeper water.

Some dolphins do closely accompany baleen whales, swimming alongside or in front of these slow-moving leviathans, and sharing in the pursuit of food, the whales taking plankton and small fry and the dolphins the fish which are feeding on these; examples are beluga with bowhead whale, and porpoises which have been photographed from

(overleaf) *Common dolphins in the Sea of Cortez, Mexico* (Topham)

the air heading in the same direction with blue and fin whales.

A retired Norwegian whale-gunner told me that he quite often saw a lone large dolphin swimming close to a blue, gray or right whale, as if for company. In his view it was an outcast from a family pod, and likely to be an old exiled male or an aged or diseased individual unable to keep up with a fast-moving school of its species; as he said, 'Waiting to die, hugging the big slow baleener for protection against the savage killer whales.'

Except for the above examples, I do not know of other instances of co-operation in the wild state for the benefit of one or both cetacean participants. At sea the species normally live separately in family pods and schools. In captivity, however, related and unrelated species in the same tank will associate and even mate, although there are no records of a fertile result. Some interesting interspecific behaviour has been observed, no doubt partly unnatural as a result of idleness and boredom, but partly derived from natural desire and reaction in a given situation.

On one occasion at Marineland, Los Angeles, a male pilot whale picked up a striped dolphin which had died in a tank containing three performing pilot whales and three striped dolphins. It is rare for a male cetacean to show such deep interest in another species, but this pilot whale was obviously distressed; its eyes opened to approximately twice normal width as it carried the corpse around the tank, holding it tenderly. It rejected several attempts to remove the dead dolphin, even plucking out a harpoon and snapping a line which was struck into the corpse. It was growing exhausted in this battle for possession between whale and man, and had to surface more often to breathe, leaving the body at the bottom. When at last the corpse was again speared and retrieved—while the whale was panting at the surface—its guardian went into a great flurry at the surface, shrilly protesting. Both he and one of his female pilot whale companions refused to perform that day, and sulked for twenty-four hours.

At the same place, a male false killer whale *Pseudorca* showed great interest in a common dolphin that had been unwell; and the trouble was apparent when she gave birth to a stillborn foetus. Attendant dolphins lifted it to the surface, but the false killer intervened, and took possession, carrying it to and from the surface for thirty-eight minutes. The mother did not seem very interested in this behaviour—perhaps she thought her baby was still alive and being cared for. But when the *Pseudorca* suddenly swallowed the cadaver, she uttered whistles of distress and swam agitatedly and rapidly around the tank.

What was the *Pseudorca's* motive in swallowing the dead baby? Hardly hunger—the cetaceans at Marineland are very adequately fed. Possibly the false killer had decided the safest and only place for it was out of sight in its stomach. Undoubtedly this large intelligent all-black dolphin had become deeply attached to the small common and striped dolphins living in the same tank. When a routine erysipelas vaccination

A pilot whale and a white-sided dolphin in captivity (John Mason/Ardea Photographics)

for its inhabitants made it necessary for the water level to be reduced to facilitate capture for this purpose, a female *Pseudorca* would not allow the men to touch one common dolphin. Each time they tried to grasp it, the false killer pushed them aside. When they persisted the *Pseudorca* grabbed one trainer's leg. She could have crushed it, but instead held it gently but firmly until the dolphin was freed. This occurred twice, with no harm to the humans, the whale was simply warning them to let the dolphin alone. The tank had to be drained further to reduce the free movement of the *Pseudorca*, but as the dolphin was being removed the false killer flung herself towards her disappearing companion and stranded on the shallow edge of the tank.

Leaping killer whale at Seaworld, San Diego (Kenneth W. Fink/Ardea Photographics)

The whale-hunters

Calling up all our reserves, we hauled up to him, regardless of pain or weariness. The skipper and mate lost no opportunities of lancing, once they were alongside, but worked like heroes, until a final plunging of the fast-dying leviathan warned us to retreat. Up he went out of the glittering foam into the upper darkness. Green columns of water arose on either side of the descending mass . . . until all was still again, except the strange low surge made as they broke over the bank of flesh passively obstructing their free sweep.

Frank Bullen, *The Cruise of the Cachalot*

The cetaceans were naturally abundant in all oceans before the era of exploitation in sailing ships. Strandings were more frequent in those bygone centuries, for a maximum population of any animal yields a correspondingly large number of casualties of unthrifty, diseased and aged individuals.

Our coast-dwelling stone-age ancestors must have been overjoyed when they found a whale or large dolphin dying or dead upon a beach. The rich flesh and fat provided a feast over many days. The oily blubber would burn, providing heat and light. The bone was useful for making primitive tools. The sharp teeth of dolphins and whales were good for cutting and could be trimmed to make fish hooks and weapons.

As an ornament the large tooth of a sperm whale was much prized. It was the centre-piece of a necklace of smaller teeth (dolphin and shark) worn by Polynesian people, including the moa-hunters who settled in New Zealand about AD 750.

From time immemorial the cetaceans and seals of the outermost coasts ever inhabited by man have been hunted. But in the midnight summer sun countries of Greenland, the Canadian Arctic, Alaska and Siberia, they provided a large part of the means of existence for the Eskimo people. Bones from middens in Alaska prove that whales were being hunted 3,400 years ago. In Norway whale drawings on rocks have been dated back to 2200 BC. The Indians of British Columbia have used the whale as a motif in their works of art, especially totem poles, as far back as we know.

The Eskimo hunted the beluga, the narwhal and the great bowhead whales in their canoes, built with traditional skill of bone and driftwood framework, covered with waterproof walrus or seal skin lashed to the

A beluga or white whale; an engraving from An Account of the Arctic Regions with a History and Description of the Northern Whale-Fishery *by William Scoresby*

frame with thongs of animal sinew. In a single-seater kayak, or the larger family-size umiak, these whales were silently approached until the ivory or wooden harpoon with its sharp slate or stone head could be darted. Attached to the harpoon was a strong line of plaited sinew and a float made of an inflated animal bladder. Each time this buoy rose to the surface as the wounded whale came up to breathe, the hunter closed in and attacked with more harpoons and the long lance.

The gray whale, calving in the warm bays of California, and the humpback breeding in the Hawaiian chain, migrate north through the Aleutians to join the bowhead whale in or near the Bering Strait. On their passage along the coasts and sounds of Western United States, British Columbia and Queen Charlotte Islands they were hunted by Haida and other Indians in large dugout canoes of cedar and fir about thirty-five feet long and paddled by eight men. The heavy yew-wood harpoon or lance was pointed with the sharp edge of a mussel or abalone shell.

Plaited cedar bark made an efficient line to join harpoon to sealskin buoy. As soon as the first lance was darted into a whale the animal was confused—as the hunters believed—by loud triumphant war-whoops. More canoes converged upon their prey, 'launching their instruments of torture and, like hounds, worrying the last life-blood from their vitals'. Each canoe claimed a share of the prize, according to the number of marked harpoons remaining in the whale when it died and was towed ashore.

The inhabitants of the long oceanic chain of the Aleutian Islands

Stylised Haida picture of a killer whale from Queen Charlotte Islands, British Columbia. Whales were hunted by these Indians who were fine seamen and used large cedarwood canoes (Ronald M. Lockley)

employed a different method, without line or float. They coated the harpoon point with poison. Whales of any species or size were approached in a frail one- or two-man kayak. As soon as one or two harpoons struck a whale, the canoe was hurriedly paddled out of danger, back to the shore. Hopefully the whale died from the injected poison within a day or two and—still hopefully—its carcase, swollen with the gases of decomposition, floated to the surface and could be towed ashore, and claimed by the owner of the harpoon.

This wasteful, almost casual form of whale-killing, was nevertheless important to the Aleuts and had little or no effect on whale numbers. The poison used was from the wild aconite or monkshood plant, which is common in the Aleutian and Pribilof Islands. Its lethal effect was well known to hunters on the nearby coasts of the Kurile Islands, Kamchatka and Hokkaido (northern Japan) where the plant flourishes. For centuries the recipe for preparing the poison was a well-kept secret among these hardy small boat whale-hunters. As Pinart naively wrote in 1872, the curious recipe he had obtained from these hunters was virtually impossible to make use of. He had been told that 'the lances are dipped before they are used, in human grease, which has been prepared from the corpses of rich persons whom the whalers have exhumed and put to boil'.

In the warmer Japanese islands south of Hokkaido whales have ever been an important part of the economy of the coast-dwelling people, hunted for centuries from rowing and sailing boats. About 1600 an unusual method of taking large whales was developed, probably as the

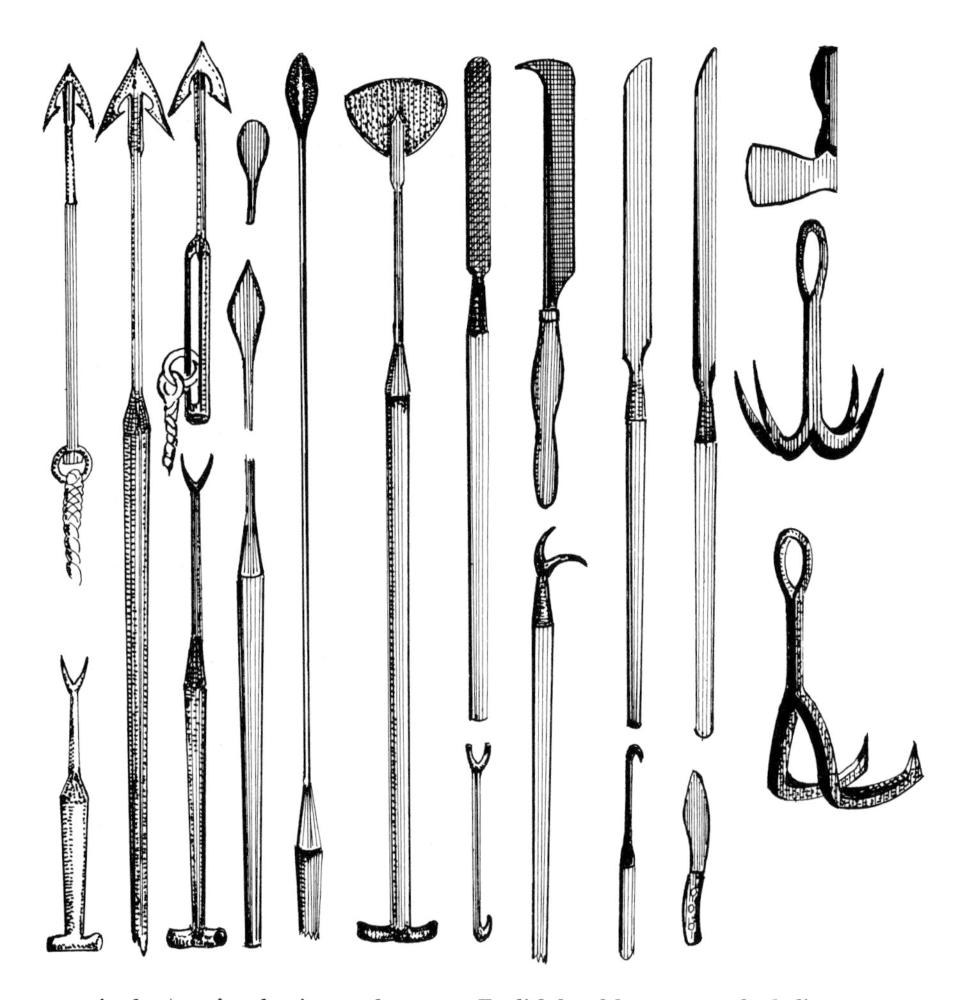

A selection of early nineteenth-century English hand harpoons and whaling gear and (opposite) *Japanese and Eskimo equipment*

result of the accidental entanglement of a young humpback in a fishing net. It is described in detail by Yosei Oyamada in his book *Illustrations of Whaling* published in 1829. Coloured plates show how the nets were set, and how divers fearlessly swam among entangled humpback whales to lance them to death. The captures were brought to a factory built at great expense by the rich merchant Masutomi on the little island of Ikitsuki. Hundreds of workmen were employed.

From the ancient history of whaling in the Pacific we return to the development of the industry along the Atlantic coast of North America. Here the native Red Indian tribes had hunted the small cetaceans and seals from their canoes with no great effect on their numbers. Strandings were still frequent on this whale-haunted coast of New England at the time of its settlement by the first colonists from the British Isles. It was

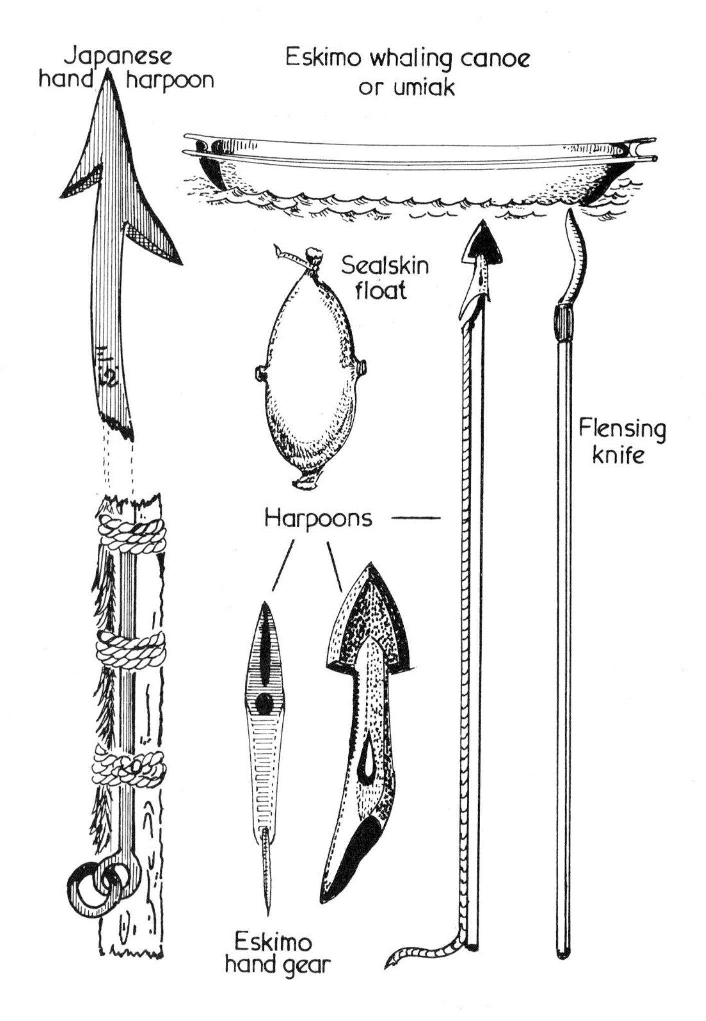

not long before local byelaws were made to regulate the disposal of the valuable meat and bone: so much to the finder, the balance to be shared by the community.

These regulations were echoes of similar laws which had long been in force on those European coasts from which Columbus and successive adventurers had sailed to discover and exploit the Americas. Thus the records of the Biscayan port of Lequitio for 1381 state that by order of the authorities the proceeds from the sale of whalebone shall be divided into three parts, two towards the repair of the harbour and one-third towards the repair of the church. Basque sailors were well-known for their skill in taking the large right whale which at one time was numerous in the turbulent seas of the Bay of Biscay, the nursery of vast shoals of immature sardines on which these whales fed. The Basque ports of Fuentarrabia, Biarritz and Bermeo still feature a whale on their coat of

The whaling ship Esk of Whitby *damaged by ice, from* An Account of the Arctic Regions . . . *by William Scoresby*

arms. As far back as the tenth century watchers were stationed on headlands and hills near the shore where often a lookout tower or a bear-pole was erected for the better view of the man on duty to signal the spouting of whales, as well as signs of shoals of fish, and no doubt of human enemies.

Still earlier when the Vikings colonised Iceland and the Faroes, they brought their knowledge of how to drive a school of pilot whales ashore, by encircling them with small boats and making much noise.

The whales frequenting the Portuguese, Spanish and French Atlantic coasts were already overhunted and scarce by the sixteenth century. Meanwhile their hardy fishermen from 1545 onwards were sailing across the Atlantic to exploit the fabulously rich Newfoundland Banks, teeming with fish, seals and whales. It was the abundant cod that made their long voyages so profitable. Salted down, it had become a staple item of diet and trade at home.

It was an age of discovery, prompted by the Spanish conquests in America. British merchants established the Muskovy Company by

Royal Charter in 1555, describing themselves as 'Merchant Adventurers of England for the Discovery of Lands, Territories, Isles, Dominions and Seignories'.

In 1578 Anthony Parkhurst from Bristol reported an average of '150 sail of French and Bretons, fifty English, fifty Portuguese, a hundred Spanish craft, all taking cod, but in addition, there were thirty to forty Spanish sail hunting the big whales,' off Newfoundland.

News of this great activity in the north-western Atlantic encouraged the search for a north-west passage to the East Indies. In 1585 the 50-ton British barque *Sunshine* sailed north of Newfoundland and west of Greenland and penetrated the Davis Strait, named after her captain; he found no passage but reported large numbers of great whales. In 1596 two Dutch ships attempted to find a north-east Arctic route to the Indies; they discovered first Bear Island, and then, across the sea named after the pilot William Barents, sighted Spitzbergen and its ice-fields. These cool unknown seas were pink with plankton, the food of numerous giant whales, and on land and ice there were seals, walrus and polar bear.

The news of Arctic waters teeming with great whales whose blubber oil had suddenly become highly profitable on the world market inspired the Muscovy Company to fit out two ships which sailed in 1610, with experienced Basque harpooners aboard, to hunt the whales of East Greenland, as Spitzbergen was known to the merchants. They returned with a profitable, if highly odorous, cargo of casks of raw oil. Containing blood and other impurities this oil had deteriorated on the long sailing voyage. In future it was to be sterilised and purified by boiling the oil out of the blubber in trypots ashore, 'the first oyle that was ever made in Greenland'.

The Muskovy Company held from Queen Elizabeth a monopoly of the Spitzbergen whale fishery; and tried to maintain it at the point of the gun when Dutch and other whaling ships arrived to share in the rich harvest of the Arctic waters. They too set up flensing stations ashore. So profitable had it become that rival English companies defied the monopoly and collaborated with the Dutch in sharing new facilities ashore. In the next five years a whole town sprang up on Spitzbergen, known as Smeerenburg (Blubbertown), with a summer population of 1,500–1,800 men attending some 300 sail of many nations working or anchored offshore.

The Muskovy Company, refusing to co-operate, and becoming inefficient in its operations, was wound up in 1615. Smeerenburg lasted more than fifty summers; the Dutch even built a church there.

The baleen giants of these Arctic seas, the blue, Greenland, and fin whales, browsing peacefully on the pink pastures, had rarely or never encountered the lofty, silent ships before, and with no natural enemies so large (towering like the icebergs they knew as harmless) they were easily approached. But the fast-moving blue and fin whales were generally too powerful to hold, breaking harpoons or snapping the line. The slow-moving Greenland whale was the mainstay of the Spitzbergen fleet. As it became scarcer profits declined, and gradually the shore stations were abandoned. Bigger sailing ships were built to pursue the whales in the open sea, north of Iceland as far as the frozen east coast of Greenland.

Whaling became pelagic. Each whaling barque carried four or more rowing-sailing boats which were launched as soon as the masthead look-out called 'Tha'ar she blows!' Many were the losses of men, boats and gear, and many a harpooned whale broke free, to die of suppurating wounds. But it was the excitement of the chase as well as the hope of a share in the profits that spurred men on. No wages were paid to the crew; they were on shares, beginning with a basic 250th part of the profit of the voyage for the lowest-paid—the apprentice and greenhorn sailor.

The dead whale was towed to the ship and chained alongside. First of all the long blanket of the blubber was cut and unrolled from the carcase

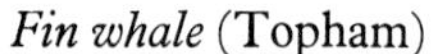

Fin whale (Topham)

The dangers of the whale fishery, from An Account of the Arctic Regions . . . *by William Scoresby*

in one continuous strip, hoisted on deck by block and tackle, sliced and fed into the trypots or boilers housed in a bricked, fireproof enclosure in the centre of the main deck. Once the oil had been boiled out of the blubber and stowed in casks the residue from the trypots burned well as fuel to keep these boilers at work.

The yield from seven large whales filled the hold of a 200-ton ship. The average catch of a Dutch ship during the five-month summer voyage in the Arctic at this early period was four to five whales. Jan Mayen, the remote island between Spitzbergen and East Greenland, was occupied for a while by the Dutch, who left it as it is today—a graveyard of whale bones and rusted trypots.

The continued demand for and the high price of whale products landed in England by the competent Dutch led to another English merchant company being formed, with the grandiose name of the South Seas Company, and the avowed intention to engage in world-wide trade. This company fitted out twelve new ships in 1725 which sailed, not to the South Seas, but to hunt whales in the Arctic. They failed to capture a single whale; tantalisingly the surviving schools were living among the ice-fields.

New heavier ships were built with double planking to enter and withstand the ice. The English government exempted from duty the small quantity of oil they brought home, and by 1740 was giving the company a bounty of 40 shillings per ton. Hull and Dundee became the chief whaling ports. By the summer of 1788 there were 347 British whaling ships at the ice limits of Greenland seas. Several were trapped by the autumn ice and crushed; others were never heard of again.

Life aboard a British whaler was often grim, and not seldom brutish. The diet was wretched: mouldy biscuit, salt meat and drinking water from casks which had been used to store oil and blubber. At long intervals there was fresh meat from fish, seals, walrus and sea-birds (the flightless great auk, penguin of the north, was slaughtered in great numbers and became extinct on the remote islets of the sub-Arctic). For some reason or other British seamen seldon ate whale meat—perhaps the conspicuous infestation of intestinal worms from which whales suffer put them off? But other European whaling crews relished fresh whale as a tasty addition to their shipboard diet.

British whalemen were exempt from the savage attentions of the press-gangs which lawfully seized and forced unwilling youths and men to serve aboard naval and merchant ships of this period.

Such was the intensity of the search and slaughter that the era of Arctic whaling was over by the mid-nineteenth century. There were too few whales left to be worth the expense of outfitting.

Meanwhile Captain James Cook had been sent to investigate the mythical Great Southern Continent which had excited great public interest and speculation. In his intrepid well-ordered voyages begun in 1768, ostensibly to observe the transit of Venus at Tahiti, he charted the coasts of New Zealand and Australia, but found only a vast whale-haunted sea around the frozen continent of Antarctica. On the last voyage in 1779, when he was slain in the Hawaiian Islands, his ships, vainly seeking a passage through the Bering Straits to Hudson's Bay, found more whales.

Whale oil for lighting and heating was in even greater demand as world populations rapidly multiplied: fossil petroleum fuels were still almost unknown.

The news of a whale-rich Pacific led to the dispatch in 1787 of four British ships, which sailed around Cape Horn to hunt whales in the cold plankton-rich waters of the Humboldt Current along the west coast of South America. Four years later six New England whalers followed.

The white settlements of the eastern seaboard of North America were now well over a hundred years old. As already mentioned, the sea-going inhabitants had long exploited the 'drift' whales, as the stranded ones were called. Following the example of the European—chiefly Basque and Portuguese—cod-fishers on the nearby Newfoundland Banks, they

A fin whale ashore at a Scandinavian whaling station in 1896 (Topham)

had hunted the migratory whales, from small boats at first. A seasonal watch was maintained on headlands. In 1650 New England ports began to build small fast barques of about 20 tons burthen, which were held in readiness to sail as soon as the signals were hoisted.

In an oil-hungry world the profits were such that the governing power (Britain) could impose substantial taxes on the rising new industry, yet another of those levies which were to exasperate the enterprising settlers and lead to 'the Boston tea party' revolt on this coast, and to the American War of Independence.

In the year 1712 Nantucket skipper Christopher Hussey was vainly cruising for right whales when a gale blew him far from land. He fell in with a school of sperm whales and, as a venture, harpooned and killed one. It proved rich in oil and spermaceti. Little did he realise that he had started an industry that was to make the New England ports the wealthiest in America for over a century to come, growing to a zenith by 1850.

The immortal story of *Moby Dick*, the novel by Herman Melville, with its stirring word-pictures of the long whaling voyage of Captain Ahab, seeking revenge upon the monstrous white cachalot which, on his first attempt to harpoon it, had thrown him into the sea and severed half one leg, is based on the author's own experience of nineteenth-century whaling. Many more strictly factual, unadorned accounts of similar adventures and disasters are to be found in the numerous log-books and diaries of whaling captains, preserved in the libraries and oceanographical institutions of their home ports in New England. One can spend absorbing hours browsing through these.

The route around the world taken by Ahab in search of Moby Dick was substantially that developed by these hardy young New England captains. Theirs was an Odyssean voyage from summer to summer, lasting two or three years. Stricter discipline was enforced than in British ships, and generally the crew worked all the better for it.

The Yankee whaler, equipped and provisioned to be independent of foreign ports of call for all save water and fresh fruit, would leave its New England base in September, crossing the Atlantic on a southeast course for the Azores and the Cape Verde Islands. It would pick up on the way any sperm or baleen whales (or if short of these, any large dolphins) the ship encountered. In the Atlantic Islands the skipper would take on volunteer Portuguese seamen, traditionally expert in pursuing whales in small boats.

Following the trade winds around the world the Yankee ship found killer, sperm and humpback whales in the cool Benguela Current off south-west Africa, and around the Cape of Good Hope to Durban and Madagascar. On then to the Pacific, depending on how fast the voyage had been so far. It was now summer in the southern hemisphere; if in good time the ship cruised south of Australia where humpback, sperm and right whales were numerous, and so to New Zealand waters. Or, if there had been too long a delay in hunting in the Indian Ocean, the whaler cruised direct through the equatorial islands—the Java Strait, past Borneo and the Philippines towards Japan.

The southern autumn—March to May—was a good hunting time. The whales were very fat from feeding on the krill and fish of the Antarctic and sub-Antarctic seas, and were migrating north to winter quarters south of the equator.

North of the equator the same species (but discrete unrelated populations which never mixed) were already moving from their subtropical winter quarters towards the Japan Sea and the vast land-bound, winter-frozen Okhotsk Ocean. Here the Yankee whaling flotillas made their rendezvous, often arriving before the Okhotsk had melted enough to allow hunting. While waiting for the right conditions, the New England ships would anchor or sail close to each other, enjoying the opportunity for a 'gam', that convivial exchange of talk and, most important, news from home, over dinner aboard ship.

In due course these plankton-rich seas were depleted by overhunting,

A bowhead (Greenland) whale with its cub and a narwhal from An Account of the Arctic Regions . . . *by William Scoresby*

and whales had to be sought as far north as the ship dare be sailed, even through the narrow straits leading to the Arctic Ocean. The Bering Strait was rarely free of ice until late July, but here lived the last of the Greenland whales which had long been exterminated in the Atlantic. The Yankees called them bowhead, from their enormous bow-shaped jaws.

What tales these skippers brought back to their families at home in New England! Yarns of wild storms, boats wrecked by great whales and ice pressure, men lost overboard; and, by contrast, days of calm, peace and feasting with the happy brown people of warm Polynesian islands where the whaler called for water and fresh fruit. Here, if a dissatisfied crew man had deserted, he could be replaced with a Kanook—as the Yankee dubbed the Polynesian—who wished to see more of the world abroad the white man's splendid ship.

Many a high-spirited young American ran away from home, lured by these tales of adventure and prospects of a fortune, an eager spirit who rose to become a captain and part-owner in a ship before he was thirty. After his first or second successful voyage he would marry. On his next return, a year or two later, he would find he had become a father. Some lonely and bored wives longed to be with their husbands aboard a fine sailing ship, enjoying his adventurous voyages. After bearing one or two children at home the more courageous might persuade their husbands to take them on subsequent cruises, leaving their children in the care of relatives.

A few of these enterprising wives kept diaries of their experiences. *One Whaling Family* is the illuminating record of the Williams family of New Bedford, Massachusetts. It begins with the highly readable journal of Eliza Azelia (née Griswold) who sailed with her husband Thomas Williams in *The Florida* on 7 September 1858. The cruise was to last over three years, and Eliza was very seasick at first. She records the birth of her third child, William Fish Williams, on 12 January 1859 after a violent gale in the Tasman Sea west of New Zealand. The next child, Mary Watkins Williams, was born aboard *The Florida* on 27 February 1861, off the coast of Mexico, on the same voyage in search of gray whales.

Eliza Williams's diary tells of great storms and blessed calms, of happy days at anchor within the coral reefs of Polynesian islands and in ports in New Zealand, Japan and Russia, of the hardships of ice and snow, and the excessive heat of the tropics. She admires, but fears, the result of her husband's skill and daring in lowering away one of the rowing boats to be among the first of his crew to tackle a whale. Her two infants thrived throughout. Born at sea, they sailed with their parents on subsequent voyages, surviving two wrecks of their father's ships: the *Hibernia* in 1870 and the *Monticello*, which was one of thirty-one whalers caught in the fast ice of the Arctic Sea in 1871. The latter disaster, when the captains and crews of these doomed ships had to work their way in small boats along the icebound coasts and escape through the Bering Straits before winter overwhelmed all, is vividly described by William Fish Williams in another part of this memorable book.

The Florida, above mentioned, was full-rigged, a fast sailor, 37.4m (123ft) long, 9.4m (30ft 10in) beam, and 4.56m (15ft) depth of hold. She had been adapted from a freight ship built in 1821. The new whaling ships were more solidly built, to withstand gales and ice, with double-planked hulls, caulked with tarred oakum, and copper-sheathed below the water line. They were about 350 tons displacement, bluff-bowed, with bowsprit high above the water, not at all fast sailers, but most seaworthy.

Yankee pelagic whaling in sailing ships reached its peak about 1850 with this exploitation of the north-west Pacific grounds—the Japan, Okhotsk and Bering Seas during July and August and in September a cautious penetration of the Arctic Ocean. In 1851 more than 700 whaleships produced 428,000 barrels of oil.

Sixty years later—1909–10 season—world whaling, now almost entirely in steam-driven ships, yielded only 340,000 barrels. Two things had happened: the whales had become very scarce, and vast reservoirs of fossil fuel underground had been discovered and were being exploited, causing a severe decline in the price of whale oil. The advent of the steam-driven whaler and processing plant also meant that fewer crew were needed to hunt the fewer whales, while much larger cargoes of whale products were carried on much shorter, one-season voyages.

Whalemen, like sailors generally, held many superstitions and curious beliefs. Certain words were unlucky, and must not be uttered aboard ship—notably the word rabbit. Women were not welcome to come aboard a ship before a voyage; for the same reason wives and daughters should not come to the quay and wave goodbye—some who did never saw their husband or father again. Eskimo whale-hunters would order their womenfolk to keep away from sight of launching the big umiak when the migrating bowhead or gray whale was about to be hunted.

Among the New England whalers of the early nineteenth century there was a belief that the harpooned and dying whale turned to the sun as it gave up its life. Even the fierce Ahab felt the strange beauty and mystery of the great sea-enemy he hated so much, yet in his heart loved for the excitement and purpose Moby Dick had brought into his hunter's life, so lyrically conveyed by Herman Melville:

> For that strange spectacle observable in all Sperm Whales dying—the turning sunwards of the head, and so expiring—that strange spectacle, beheld of such a placid evening, somehow to Ahab conveyed a wondrousness unknown before. He turns and turns him to it—how slowly, but how steadfastly, his homage-rendering and invoking brow, with his last dying motions. He too worships fire; most faithful, broad, baronial vassal of the sun!

The badge of the Discovery *expeditions*

Modern whaling

Canst thou draw out leviathan with an hook?
Who can open the doors of his face?
By his neesings a light doth shine, and his eyes
Are like the eyelids of morning.
When he raiseth up himself, the mighty are afraid.
The Book of Job

It was rare for the hand-harpooners in the rowing boats of the sailing barques to tackle the large rorqual or fin whales—the blue, fin and sei—in the open sea. They were too fast and powerful and if harpooned would dive and tow the boat dangerously and break the line; they took very much longer to kill, and when dead would sink quickly.

With the invention of the harpoon gun, mounted on the bow of a steam-driven ship, all this was changed, and the harrassment of the great whales leading to their present low numbers, near to extinction, began in the middle of the last century. At first this gun fired a harpoon with hinged barbs. This entered the whale's flesh and as soon as the strain was taken up on the attached line by the diving whale and the ship's winch, the flukes flew open, giving a firm purchase on the quarry. Even so, the blue and fin whales easily snapped the heaviest line as they surged away in the deep dive, later to die of suppurating wounds.

For some years a Norwegian Captain, Svend Foyn, described as 'a good religious man respected and beloved of all who met him', had been experimenting with explosives placed in the barbed head of a harpoon. At the first test of this weapon in 1864 the line caught his foot, and he went overboard with it; and in the log-book duly thanked the Almighty for saving him. In 1868 he improved the harpoon by fitting a delayed-action charge in its head. This did not explode until after the harpoon had entered the whale; as soon as the strain on the line pulled the hinged barbs open, these broke a glass phial containing sulphuric acid, thus igniting the cylinder of gunpowder which exploded deep in the body of the victim.

Subsequently a fuse replaced the glass phial, and it is this fiendish weapon, still further refined, which is responsible for the near-extermination of the great whales today. There is no longer any danger to the whale-hunter; the days of capsizing and drowning in small boats, the excitement and adventure and risk of hand-throwing the harpoon have

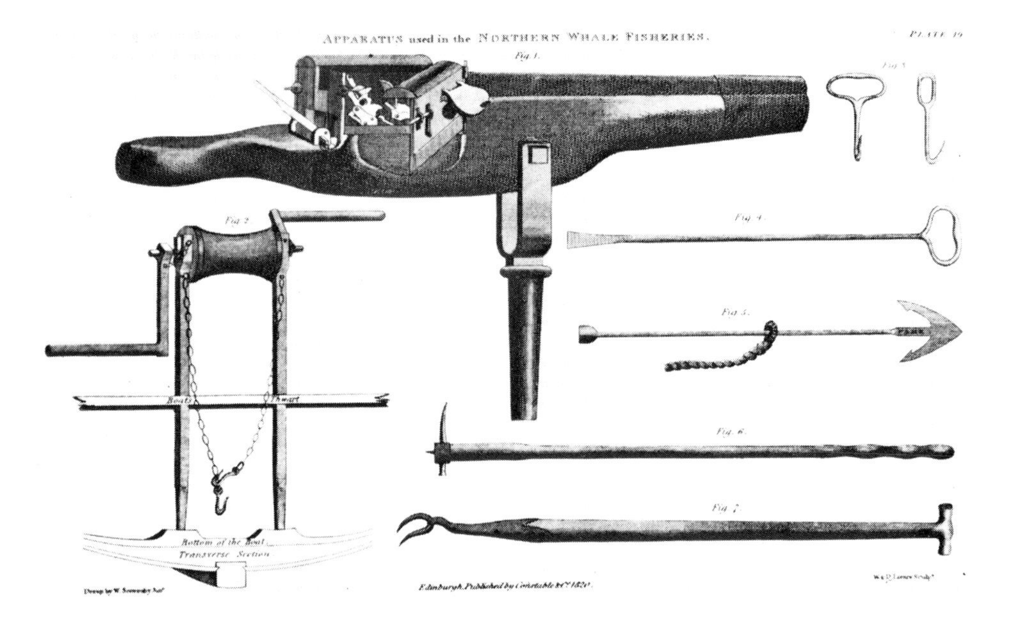

Whaling apparatus illustrated in An Account of the Arctic Regions . . . *by William Scoresby*

Svend Foyn harpoon gun

vanished, replaced with an almost monotonously safe procedure. The gunner, awaiting the old signal from the masthead lookout of 'Tha'ar she blows'—even this is largely replaced by a telephone call—is comfortably housed in his cabin amidships. He proceeds by a safe catwalk to his gun, mounted on the ship's bow.

Some skill of aim is required, of course. But the whale-chaser boat is fast, and quickly catches up with the unsuspecting animal. Even so a great whale does not die at once, unless the explosive charge reaches its brain with the first harpoon fired. If it does not the whale may swim deep and far before rising to the surface to breathe. The line must be winched in as the chaser catches up with the wounded leviathan, the gunner waiting for the right moment to fire a second shell more accurately. This may not be necessary, however, if the victim is spouting blood when it appears, the signal that it will soon begin the 'flurry', as it is called, the agonised writhing and tail-lashing of the death throes of a sentient giant, a sight to melt the heart of the most experienced whale-hunter.

'Only the thought of that little farm in the fiord my wife and I have been saving up to buy,' a retired Norwegian whale-hunter told me, 'enabled me to fire that bloody horrible gun at these poor friendly creatures. Every death flurry has given me a sleepless night or nightmare.'

The modern whale-chaser, working in conjunction with a factory ship or a shore-based station, is a small fast craft of about 120ft length, little more than a container for its powerful diesel engines, a few miles of rope, a crow's nest mast, and a platform for the gun. As soon as the gunner is called to the bows he directs the helmsman in the wheelhouse with hand signals, giving the direction and speed—starboard, port, slow, fast, stop. He is expert at assessing when and where the sighted whale will surface to breathe. A whale previously hunted and perhaps wounded will flee from the approaching chaser, and at first will outdistance it in spurts of 20 knots. But the marksman knows it must come up to breathe, and part of his reputation lies in his skill in estimating the quarry's position when it does next blow. Unaccustomed to high-speed swimming for more than a few minutes, the whale slows down, coming more frequently to the surface to blow: it becomes agitated by the long chase and 'pants' longer. Today there is small chance of it escaping, for it can be followed by the ship's sonar under the water.

Svend Foyn was also the inventor of a device which has the effect of a well-sprung fishing rod and line. The sudden savage tension on the harpoon line between diving whale and rolling, pitching ship is relieved by passing the line over a sheave mounted on strong springs attached to the foremast, springs which absorb the shock. Meanwhile the line is

Humpback-whale carcases, inflated to prevent sinking, being towed to a factory ship (Topham)

being winched home, the strain further relieved by shock absorbers elsewhere along the winch line. As soon as the dead whale is alongside, a hollow lance attached to an air pump is thrust into its body and the carcass is inflated to keep it afloat. The hole made by the lance is plugged with a stopper, soaked in anti-bird fluid to discourage its removal by the numerous sea-bird scavengers which pick at the carcase. The whale-catcher's individual mark is notched in the flukes of the tail as evidence of ownership and the carcass set adrift, to be collected later.

Foyn's inventions gave the Norwegian whale fishery a much needed superiority and return to prosperity. The fast swimming blue and fin whales could be killed by the hundreds. In the 1885 season 1,398 were slaughtered off Finnmark and processed at shore stations equipped with the latest steam-driven machinery. Net and line coast-fishermen made an outcry that the hunting of whales inshore caused so much disturbance, blood and filth that the shoals of herring and other marketable fishes had been driven away. At last in 1904 the Norwegian government banned the catching and processing of whales within territorial waters. They had in any case already become overhunted and scarce.

With the huge profits Norwegian whaling companies began to establish new bases right across the world, using faster chasers to harpoon every sort of whale they could kill with the deadly delayed action bomb-gun. They registered new whaling companies under the flags of other nations—but employed expert Norwegian gunners. As the great whales of the northern hemisphere were reduced to a point where it was no longer profitable to hunt them, those of the southern oceans came under pressure, off the coasts of South Africa, the Indian Ocean, Australia, New Zealand, Cape Horn and the Falkland Islands.

The government of the Falkland Islands, as part of the British Empire, controlled the small islands as far south as the polar ice. By 1911 it had granted eight leases for whaling stations in South Georgia. The Norwegians set up the first at Grytviken Bay, where a miniature town flourished, inhabited by Norwegian whalemen and British Excise officials.

The summer concentraction of krill whale-feed around South Georgia and the chain of the Antarctic islands (South Orkneys and Shetlands) is probably the greatest in the world. The dues imposed upon the visiting whale-ships were at first large, and paid for the services of scientists to study, measure and take samples of each whale brought to the flensing works of two stations set up in South Georgia—the second was known as Leith Harbour. This resulted in a considerable extension of our knowledge of cetacean anatomy and breeding biology.

The whale-chaser that killed early in the day would plant its company house-flag in the body of the inflated whale, and leave it to be collected later. Two to four large whales were enough for the chaser to tow back

A general view of the meat deck of the British whale factory ship Balaena, *partly obscured by boiler steam* (Topham)

to the shore station. As they killed the whales feeding near South Georgia, the chaser had to range further afield for victims, and make correspondingly longer journeys back to base. It was time to adopt the more economic practice of the latter-day sailing ships, and process the whales at sea.

In 1903 a Norwegian company sent the ship *Telegraph*, 500 tons, to be used as a processing factory, with two chasers in attendance, to hunt Spitzbergen seas. In the same year the first large pelagic factory ship, *Admirale*, 1,500 tons, visited the Falklands. Attended by fast chasers she ranged south to the Antarctic islands, taking blue, fin, humpback, sei and sperm whales. She was followed in subsequent summers by a fleet of factory ships of British and other nations working around the whole iceberg-infested coast of the Antarctic continent. Between them they slaughtered annually up to and above 30,000 whales.

Fossil fuels were continuing to flood the market and reduce the price of burning and lubricating oils and to make a profit it had become necessary to kill and make use of every saleable part of the whale.

The advantages of pelagic whaling were obvious: no customs levies at shore stations, and the factory ship followed the chasers as they pursued the migrating whales. Some firms now supplied a 'buoy boat', generally an old, slow, seaworthy tug-class ship which collected and towed to the mother ship the whales that had been killed, inflated and flagged by the chasers, leaving these free to concentrate on hunting. A radio transmitter planted with the flag pole enabled the buoy boat to locate the marked whale on its radar screen. Pelagic whaling on a commercial scale led to the closing down of many shore stations or their reduction to small-scale operation.

From 1954 onwards spotter aircraft were flown from the Durban whaling base, on the east coast of South Africa, to report back to and guide the gunboats to schools of migrating sperm and humpback whales. The final refinement for the location of whales has been the helicopter, especially on the last of the large factory ships, at present used pelagically by Russia and Japan.

The modern factory ship has a slipway in the stern for hauling the whale to the flensing deck. It is today's boast that the largest whale can be cut up and disposed of completely within one hour, where it took a whole day in the time of the sailing ship a hundred years ago.

Blubber

The main revenue is obtained from the oil with which the whole body of the whale is impregnated. The blubber or outer coat of fat, rolled off in a continuous strip, cut up and fed into boilers, yields up to 80 per cent by weight in extracted oil. The bones yield up to 60 per cent, and the meat a maximum of 7 per cent of raw weight in oil. This is a vastly improved yield on the old process of trying out aboard sailing ships, when a single whale, perhaps after lying in the sea for a day or a night

The Balaena, *operating around 1947, had three Walrus spotter planes whose pilots qualified to wear the whale insignia above their wings* (Topham)

was ravished by sea-birds and sharks and had begun to decompose, yielding stinking oil of inferior quality.

The most superior fine oil comes from the blubber of the baleen whales which have fattened over a summer feeding on the krill of polar seas. A blue whale yields between 10 and 20 tons; a record 50 tons of oil was obtained from a blue at Walvis Bay, South-West Africa, in 1928. A fin whale yields up to 8 tons of oil.

The oil was formerly stored in 40-gallon barrels for easy handling. Some of these inevitably leaked during the long sailing voyages, adding yet more oil to the wooden hulls of the early whalers, whose whole life at sea was permeated with the unpleasant odour, which nevertheless was the reek of prosperity and hardly noticed by the crew. Today oil is stored in large tanks in the factory ship.

Whale oil

Whale oil is processed into various ingredients and end products, including edible fats (margarine and frying fat), soap, tallow, glycerine

Two whalemen demonstrating the size of a sperm whale's jaw (Topham)

(the raw material of high explosive nitro-glycerine), machine oil, fine grease (for textile thread conditioning), tanning, waterproofing and chamois leather, paint oils. As most of these products are today also processed from fish and vegetable and petroleum oils, and some synthetically from raw chemicals, the price of whale oil on the world market varies accordingly.

It is somewhat ironic that conservation of whale stocks was not seriously considered by the companies who had hunted the whales to near extinction until the price slumped from around £40 a ton a hundred years ago to a bare £10 in 1930. Alas, it was not love of whales, but loss of profit that brought about the present international agreement to restrict future catches.

In chemical terms the composition of baleen-whale oil is an ester of acidic fat containing trihydric alcohol-glycerine. That of sperm-whale oil is a liquid wax, esters of fat with monohydric alcohol. This waxy oil was formerly in great demand for making candles, since it burns with a white smokeless flame. Today it provides a base for pastes and unguents used in cosmetics.

Spermaceti

The enormous head of the adult sperm whale, weighing several tons, contains about one ton of spermaceti, a waxy oil which solidifies rapidly at low termperatures. (The terms 'sperm whale' and 'spermaceti' derive from the curious belief of those who named this whale that it carried its semen in its head!) It is now almost certain that this rich oil, stored in the 'case' (known as melon today), a network of oil-filled communicating cavities in front of the brain, is part of its sonar system. Spermaceti oil continues to fetch a high price today for its special qualities in the preparation of cosmetics, lubricating oil and scouring agents.

Whale meat

Baleen-whale meat is more palatable than that of the sperm and other toothed cetaceans: both have a strong flavour, needing some patience to

Example of scrimshaw carved on the tooth of a sperm whale

acquire a liking for. The sophisticated British public were reluctant to eat it as a substitute for domestic animal meat, advocated by the Ministry of Food during the wartime shortage of protein, and most people who tasted it fed it to dog or cat. But whale meat, properly prepared, is considered a delicacy and is in demand in other countries, especially in the Mediterranean and the Far East. The body meat, separated from the blubber, contains 21 per cent protein, and barely 8 per cent fat. It is much richer than beef in histidine, so necessary for growth.

Tail meat contains more fat: the Japanese like to eat this portion, flavoured with soy and ginger, for which purpose it is quick-frozen for the market. Nothing of the cetacean is wasted in Japan: the less palatable flesh is processed as raw material for pies and sausages.

In Iceland and the Faroe Islands the meat of pilot whales, taken by driving a school ashore, is hung raw in strips to cure in the wind and sun and during rain under cover of a store shed with slatted walls to admit the wind. I have eaten this dried black meat raw—a bit rank but good enough with appetite; prepared by sousing in vinegar and cooking it can be delicious—a popular local dish called rengi.

The skin of the narwhal is eaten raw by the Eskimo, and is rich in anti-scorbutic vitamins.

Other by-products

Other products extracted today from whales flensed within a few hours of killing include glue and gelatin from the tissues, vitamin A from the liver, insulin from the pancreas, ACTH (adreno-cortico-tropic hormone) from the pituitary (used in treatment of arthritis), meat and bone-meal for animal feeding-stuffs and fertilizer. Little is wasted today after the main product, oil, has been extracted.

Bone and Tooth

The giant lower jawbone of the baleen whale, firmly planted upright in the ground, made a solid everlasting arch to support the roof of the home of the Eskimo and other hunters and scavengers of whales on northern coasts. Bone and teeth provided material for their weapons, tools and household utensils, including needles and bodkins for sewing, and were also shaped and scraped into fish-hooks.

Scrimshaw is the name given to the designs and decorations inscribed upon or carved out of whale bone and whale teeth, including narwhal tusk. The art was much practised by whale-men, using knife, file and polisher, in the idle hours of months spent in sailing ships at sea, and many beautiful examples may be seen in museums and homes in the old whaling ports. It is an immemorial art practised by Eskimo coastal tribes, who have continued to this day to carve and decorate whale,

An orca or killer whale (Bruce Coleman)
A pod of killer whales (Jen & Des Bartlett/Bruce Coleman)

seal and walrus bone and tusk ivory, also the heavy soapstone, producing with skill miniatures of familiar animals and objects of their daily lives. These ornaments find a ready sale to tourists.

Baleen

Baleen is the remarkable horny growth of plates each side of the upper jaw of the toothless whales. These replace teeth, but grow continuously, like human nails, and are composed of a similar substance, wearing away at the tip. There are up to 350 of these plates, hollow tubes which curve inwards, each triangular in section, a few millimetres apart, fringed with an interlocking web of tough hair or bristles. As described earlier, the baleen whale sucks in a dense mass of plankton, then strains the water through the baleen sieve with its huge tongue before swallowing the mouthful of up to one ton of krill. The plates are longest where they grow from the middle of each side of the upper jaw, often more than 3m (10ft) long in the blue whale, and progressively shorter towards the hinge and the tip of each jaw.

Because of their flexible strength these plates were of great commercial value for stiffening corsets and were more durable than cane. The bristle or hair was first removed and used in making wigs, helmet plumes, shoulder pads, and stuffing furniture. Today, with baleen whales given total protection, the plates and bristles are replaced with steel and plastic and other animal-bristle and man-made fibres.

Ambergris

A product well known to the ancients, and in China regarded as a tonic and aphrodisiac when incorporated in a food dish. As found in the lower intestine of the sperm whale it is a dark stinking accretion, its source unknown. When voided it crystallizes and becomes sweet smelling and, after long weathering, assumes the colour of amber. In this condition it was sometimes mistaken for the golden-coloured fossil resin of coniferous trees, known as amber, hence the name ambergris.

Occasionally a cachalot has been unable to void a larger concretion. The record lump of ambergris weighing 451kg (1,003lb) was found in the gut of a sperm killed in Australian waters in 1912 by a small nearly bankrupt whaling company. It was sold for £23,000, a huge sum at that time, but less than half its current value. It saved the company during a poor whaling season. Today the very limited supply of ambergris is in great demand as a base for volatile essences used in high-quality perfume.

Belly of humpback whale showing grooves and genitals (Charles Jurasz)
A humpback whale showing baleen plates (Charles Jurasz)

(overleaf) *Whales were an important source of meat in the Faroe Islands; here in 1946 a school of pilot whales is driven ashore by islanders manning all available boats. On the beach and in shallow water the whales are slaughtered to provide meat, oil and bone, and sometimes valuable ambergris* (Topham)

The great whales

And God created great whales, and every living creature that moveth, which the waters brought forth abundantly, after their kind. And God blessed them, saying, Be fruitful and multiply, and fill the waters in the seas.
Genesis I

The full list of the living cetaceans of the world is given on pages 192-4. Length is measured from tip of snout in a straight line to the middle of the tail, which usually has a median notch; weight is in kilograms or tonnes: 1kg=2.2lb; 1 metric tonne is only 18kg (40lb) less than the English ton. Estimates of past and present stocks of hunted species are given on page 187.

BALEEN WHALES

Blue Whale *Balaenoptera musculus*

The largest known animal ever to inhabit the world. The maximum length is 33.5m (110ft) and weight up to 140 tonnes. In the upper jaw, 270 to 395 pairs of baleen plates varying from 59 to 104cm (23 to 41in) in length. The female in all baleen whales is slightly larger than the male. There are up to 118 grooves or pleats extending from the chin to the navel.

At sea this whale surfaces with a blast 6-10m (20-33ft) high, then the long back appears, ending in a small dorsal-fin as it submerges, rarely exposing the tail. It is a fast swimmer and was too powerful and heavy for the whale-ships of the sailing days, snapping and swimming away with the harpoon line on the occasions when it was struck—until the bomb gun was invented which exploded a lethal charge in its body. This weapon, and fast chasers, resulted in its near-extinction. Three discrete populations are known to exist, which feed in summer respectively in Antarctic seas, in the north Atlantic, and the north Pacific as far as the pack-ice limit. Its fate is still in the balance.

The blue is the first whale to return to polar seas in the spring when, like other migratory whales, it is lean from calving and breeding activities and, it is believed, little feeding over the winter period spent in tropical seas. It is then often yellowish beneath, caused by a film of diatoms (acquired in warm water) which, contrasting with the dark back,

A whale's jawbone being processed aboard the factory ship Balaena (Topham)

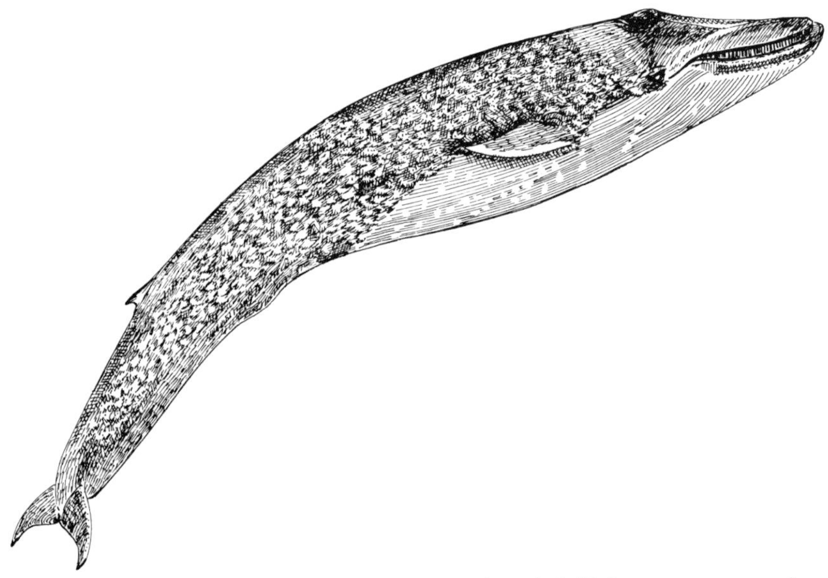

Blue whale Balaenoptera musculus

earned it the name 'sulphurbottom' among the whaling fraternity. It puts on weight with incredible rapidity as it feeds on the huge swarms of polar krill; up to one ton has been measured in one stomach cut open on a flensing platform.

This whale has not often been seen during the winter; its breeding biology has been unravelled from measurements and post-mortem examination at whaling stations. The gestation period is just under one year; at birth in the tropics the calf (rarely twins) is estimated to weigh over 5 tonnes and measures 7m (23ft) long. For the first six months, feeding exclusively on its mother's milk, it lengthens at the rate of about 4cm (1⅞in) a day, and on average puts on some 90kg (200lb) every 24 hours; a prodigious strain on the mother, as might be supposed, but in fact at this season she too is gaining weight as she engulfs many tons of krill daily during the continuous sunlight of the polar summer.

The calf is weaned at seven months when about 16m (52.5ft) long. The baleen plates grow with a distinct pattern of annual ridges up to seven years, enabling the immature whale to be correctly aged when killed during that period. At four years old and 22.5m (74ft) long only 6 per cent are sexually mature. Growth continues up to the tenth year by which time the blue whale is 100 per cent sexually mature and averages 30.5m (100ft) in length.

There is no information on how often mating occurs; probably annually, but after parturition the cow is unlikely to conceive during the long lactation. The male's penis at 2.9m (9.5ft) exceeds the length of the flipper, 2.3m (7.5ft). Judging by the very slow rate of recovery of population since it was afforded total protection in 1966 by the IWC, the mature female probably does not calve more often than once in three years.

This whale is now so rare that it is seldom seen except singly, or in pairs or threes. But it does not appear ever to have been seen in large pods. It is said to be shy and, not surprisingly considering its size, rarely jumps clear of the surface.

Since adult blue whales became scarce in the present century Japanese whalers claim to have discovered a new species, which they have called the pygmy blue, in the Antarctic between 50° and 80°S; this is regarded with suspicion by some cetologists of other nations, who believe it may be an excuse to take immature blue whales; but others, including Dr R. Gambell, say it is likely to be at least a good subspecies.

Fin Whale *Balaenoptera physalus*

Also known as razorback to the old whalers from the long back and prominent ridge of the tail-stock and dorsal-fin, features which breach the surface after each blow, usually without exposing the flukes of the tail. A medium-large whale, up to 25m (80ft) long, and more than 60 tonnes. Resembles in shape the blue, but paler on the back, and the underside is white. Up to 114 grooves or pleats extending from the chin to the navel, enabling the throat to expand to enclose several hundred kilograms of food and water in one mouthful. Baleen plates are as numerous and as long as in the blue. The blast is 5-7m (16-23ft) high.

This whale is rather less tropical and polar in its migrations, and is found in deeper water from the sub-tropics to the sub-polar. Its blubber is thinner at 6.5cm (2.6in) than the blue whale's 10cm (4in) and was therefore regarded less highly; but it has been sought out and slaughtered since the blue whale became rare and is no longer as plentiful as it was early in the present century when about 10,000 were killed annually.

Farley Mowat (1972) describes how the fin whale homes upon a fish school by echo-location, transmitting a pulse of low frequency which

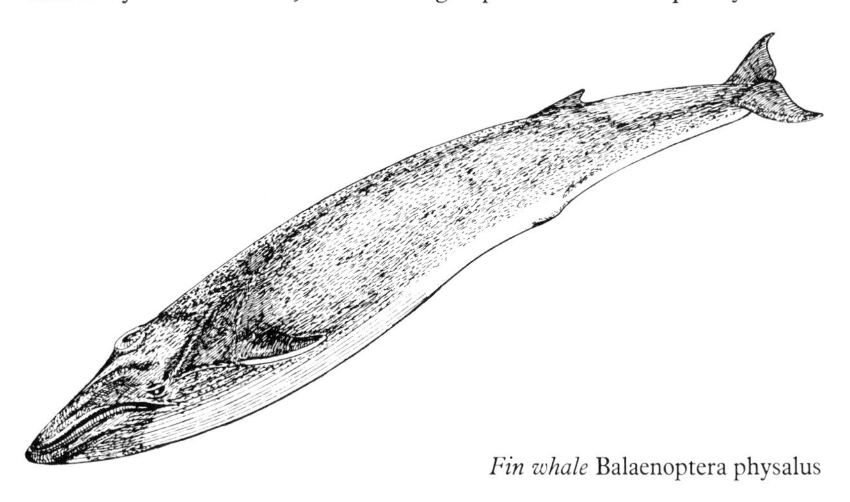

Fin whale Balaenoptera physalus

enables it to identify the species it prefers—herring. 'Closing with its target it begins circling the school at torpedo velocity of 20 knots. Now it sways sideways to present its white belly to the school, a flashing ring of reflected light,' Once the fish are concentrated in a tight bunch, scared by the white flash of the whale's underside and flippers, the finner expands the accordion of its huge pleated throat and sucks in the ball of fish. The muscles of the throat then contract as the great tongue rises to expel the surplus water through the narrow slits of the baleen plates, which retain the food for swallowing.

A similar hunting technique has been seen in humpbacks, and probably all the baleen whales round up and ingest their normal food of plankton and shoals of small fish near the surface by a circling movement, making use of sonar and reflected light. Curiously, the left side of the finner's head is dark, but the right side is pale: does this whale therefore circle its food clockwise?

The finner does not migrate nearer to the equator than approximately 30°N or S; thus the two races, isolated respectively in the north and south hemisphere oceans, have evolved into subspecies, the southern race averaging about 1.3m (4.3ft) longer than the northern.

This and the next species have been exploited from bases in South Africa, where Gambell (1968) found that there is a two-year breeding cycle of 12 months' gestation, 6 months' lactation and 6 months' anoestrus. On average 1.43 ovulations precede first pregnancy. Fin calves are 6.7m (22ft) long and weigh around 4 tonnes at birth.

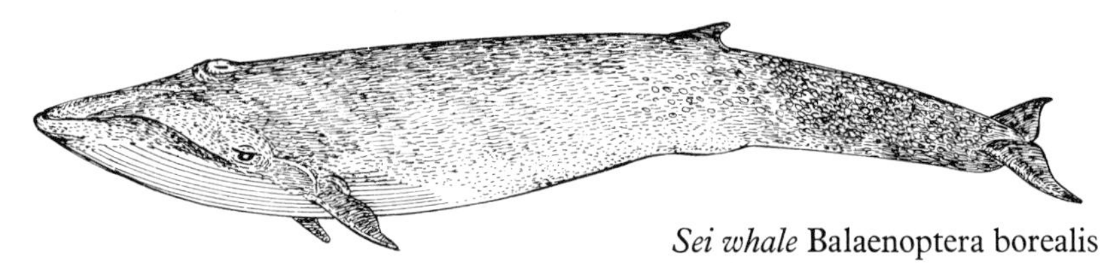

Sei whale Balaenoptera borealis

Sei Whale *Balaenoptera borealis*

A smaller edition of the fin whale, up to 18m (59ft) long and 22 tonnes, white below and dark above. Usually carries 340 pairs of baleen plates averaging 60cm (24in) in length; but the throat pleating is shorter, some 60 grooves reaching to just beyond the flipper. Much the same distribution of north and south subspecies, but does not enter the Bering Sea; its southward limit is South Georgia, where Gambell records that the blubber is thickest on pregnant and thinnest on lactating females.

Bryde's Whale *Balaenoptera edeni*

This whale is almost identical with the sei and was not separated until 1912. It is rather smaller, more slender and darker below. Up to 15m (50ft) and 17 tonnes, with 270 pairs of shorter baleen plates and some 50 throat pleats which extend to the navel. Believed chiefly to eat shoaling fish, including dogfish; occasionally birds, perhaps by accident (15 penguins were found in one stomach). Confined to tropical and subtropical seas.

Minke Whale *Balaenoptera acutorostrata*
Smallest of the baleen whales, with a conspicuous dorsal-fin. Dark above, white below. A lively plumpish whale, frequently jumping above the surface and not much afraid of ships. Length up to 9.2m (30.2ft) and 10 tonnes weight. The white oval above the flipper is absent in the southern

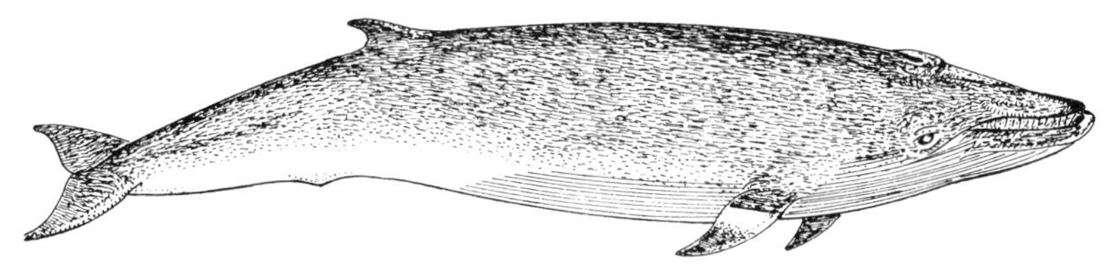

Lesser rorqual or minke whale Balaenoptera acutorostrata

colour-phase *bonaerensis*. Travels fast, and reaches the pack ice in summer, often met close inshore. The baleen plates vary between 230 and 348 pairs (fewest in the north Pacific *davidsoni* race); 50 to 70 short throat pleats. A krill, sandlace and capelin feeder. Pairing takes place in late winter in subtropical waters. Gestation is 10 months, and sexual maturity is reached at 2 years.

The scarcity of larger whales has resulted in the minke being extensively hunted and reduced in numbers. During migration the males precede the females, which move more slowly with attendant calves and weaners.

Black Right Whale *Balaena glacialis*
So-called because, swimming slowly and sluggishly, it was the right one (easiest) to kill of all the whales—also it floated when dead. It was close to extinction before it was given international protection in 1936. Up to 17.6m (58ft) long and 70 tonnes, yielding much oil from blubber up to 30cm (12in) thick. Easily identified from its purple-black colour heavily dotted with whitish concentrations of parasites (whale-lice especially), absence of dorsal-fin, horny lump known as the bonnet at the tip of the snout, and the very wide tail-flukes, measuring more than one-third the length of the body; these invariably rise vertically above the surface when this whale sounds after breaching. The double blast is 4/5m (13/16ft) high.

There are no throat pleats, but the baleen plates are numerous (up to 250 pairs) and up to 2.7m (9ft) long in this almost exclusively plankton feeder.

Almost 50 per cent of the body weight of this plump whale is fat and blubber, making it float high in the water, before and after death. As many as 20,000 of the southern race of the right whale were slaughtered in a single year late in the last century. Today the few survivors are to

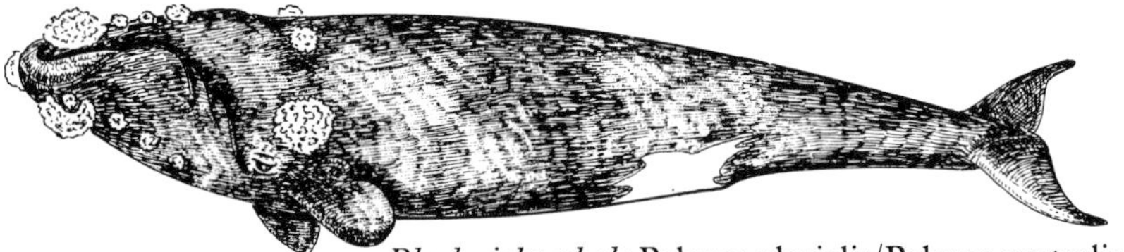

Black right whale Balaena glacialis/Baleana australis

be met with chiefly off the coasts of Patagonia, Tristan da Cunha, (South Atlantic), some sub-Antarctic islands (some 130–200 off the Campbell Islands of New Zealand), and a very few in the temperate North Atlantic. The southern race mates in August, the calves are born in July. This species does not penetrate the polar seas.

Studying and filming the behaviour of a group close to the Patagonian coast, Roger Payne records that the males were so ardent in courting that the female adopted evasive action by floating on her back or elevating her huge tail high above the surface for long periods (in this position in a strong wind the tail acts as a sail to move the whale). Intromission can only be achieved by belly to belly contact, usually with the male sliding on his back beneath the female floating horizontally in the normal position. Evidently this species is polyandrous: the males depart from this nursery coast before the females, leaving mothers and their calves to enjoy a period of peace and tender affection and childish games. At times the calf playfully flings itself on top of the maternal head, temporarily blocking the cow's nostrils.

Yehudi Menuhin's son Krov and his wife, diving among the right whales here, found them tame or indifferent to the proximity of human swimmers, who could reach out and touch the immense slow-moving bodies. Bruyns (1971) considers the genus *Balaena* (not to be confused with *Balaenoptera*) 'stout, heavy, slow-moving, stupid and defenceless'. But the experience with this southern right whale gave the Menuhins a wonderful feeling of living in harmony with a community of gentle, graceful, wise and friendly giants, and a sense of deep spiritual loss on leaving them.

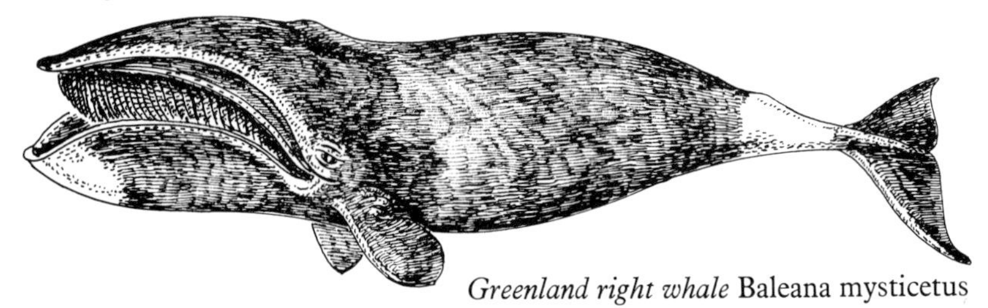

Greenland right whale Baleana mysticetus

Greenland Right or Bowhead Whale *Balaena mysticetus*

This is the strictly Arctic whale described earlier as overhunted to extinction in the high latitudes close to the pack-ice of Spitzbergen, Jan Mayen and Davis Strait in the early days of whaling under sail. It reaches a maximum length of 18.5m (61ft) and up to 100 tonnes weight. About 350 pairs of black baleen plates, no throat grooves. It was believed that the last one to be taken commercially was harpooned in 1899. The American whaling fleet of that century eagerly pursued this plump blubber-rich sluggishly moving creature as it migrated north through the Bering Strait into the Arctic Ocean with the melting of the winter ice, late in the summer; the tough double-sheathed sailing barques reached eastwards into the Beaufort Sea where they were not infrequently trapped and crushed in the autumn by the sudden onset of drifting ice and winter.

During the present century the Eskimo people of Arctic Alaska have continued to take a few bowheads each summer, by harpooning in the traditional manner but never in such quantity as to seriously reduce their already depleted population. During a visit in 1978 to St Lawrence Island, south of the Bering Strait, I found the Eskimo community of that island somewhat indignant that the IWC had fixed a quota of under twenty bowheads to be taken (by Eskimos only) for subsistence use in the whole of Alaska, since they maintained that bowheads were and always had been sufficiently numerous on migration for them to be taken as required—that is, strictly for food, oil and bone, using every bit of the whale, as the Eskimos of the Arctic have done for at least 3,000 years. One had been taken a few weeks before my arrival on this island and had been fully utilised; the immense jaws, measuring 6m (20ft) from tip to base of skull, were still being scraped of the last scraps of meat and baleen—350 pairs of plates, up to 4.5m (15ft) long; I could stand upright within these jaws, for the head of this whale measures about 6m (20ft), one-third of its body length. The blubber is 50cm (20in) thick, a warm shield against the zero temperatures of its Arctic habitat. Estimated weight of this adult was over 100 tonnes. This beast required 100 men using tackle to haul it ashore.

The bowhead in this century has become—has been driven to become—an inhabitant principally of the cold seas between the Aleutian Islands chain and the Arctic Ocean west and north of Alaska. With the recent extension of United States control of the sea for 200 miles off Alaska, there is new responsibility and interest in saving the bowhead from extinction. An aerial survey and count of the sea-mammals in the Bering Sea and Arctic Ocean in 1978 to some degree vindicated the claim made by the Eskimos; the survey estimated that a minimum of 2,300 bowhead whales exist within US (200 mile limit) waters; and as Russian sources have estimated that there are some 500 bowheads off the coast of NE Siberia, the world population of bowhead whales in 1978 was about 2,800. This gives better hope that the species will survive; the few taken by the Eskimos strictly for subsistance on an annual quota basis will not seriously affect its recovery. As one or two bowheads have in this same year been reported far to the east, between Baffin Island and Greenland, it may well be on the increase and moving gradually into its former Arctic haunts on the Atlantic side.

Living in such chill seas, this fat round-bodied whale is not troubled by external parasites. There is no dorsal-fin, the flippers are thick and long—2.4 x 1.2m (7.9 x 3.9ft), the tail-stock narrow. The tail-flukes have a spread of 8m (26ft) and are usually lifted well above the surface when the bowhead dives after breaching and sounding the loud, 6m (20ft) high blast. Pairing takes place in its southern winter quarters, in the latitude of the Aleutian chain, where the calf is born in waters still cold (maximum 5°C), and at a season when whale-hunting ships are absent.

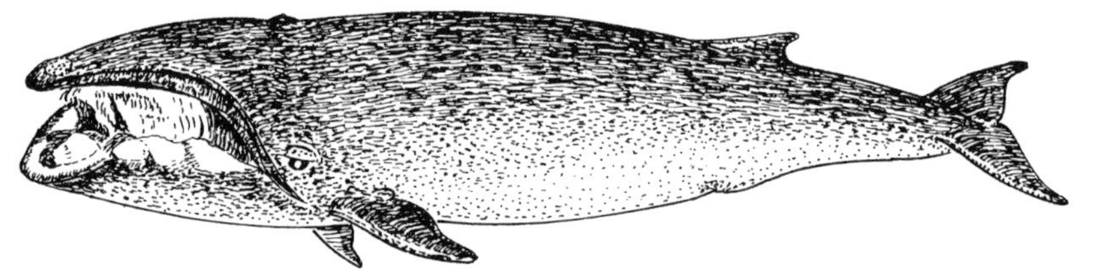

Pygmy right whale Caperea marginata

Pygmy Right Whale *Caperea marginata*

Almost nothing is known about the life history of this small right whale, one-third the size of the bowhead, which it somewhat resembles in shape and dark colour, but has a distinct dorsal-fin. It is known from less than 20 strandings in the southern hemisphere, most of them on the New Zealand coast. Typical of a plankton eater, one specimen had 230 pairs of baleen plates.

Gray Whale *Eschrichtius robustus*

A coastal species which became extinct in the North Atlantic Ocean about 1775. The population in the north Pacific Ocean was very large at that time, wintering both sides, around Japan and California, and moving north right into the Arctic Ocean as they followed the retreat of the pack-ice in summer. They were ruthlessly hunted by American and other whaling fleets which were also taking bowhead and sperm whales whenever they could. When the nurseries of the gray in warm Californian lagoons were discovered, the cows with their new-born calves were easily slaughtered; this eastern Pacific population was believed to be totally wiped out by 1911. On the western side, the gray whales wintering off Japan and entering the Okhotsk and Bering Seas in summer were reduced to their lowest numbers by 1966. Arctic Eskimos were, however, still taking them as a remnant continued to pass north or south through the Bering Strait.

A few reappeared off California in 1925 and were given total protection. In fifty years they have increased, and this eastern wintering stock is now estimated to be well above 6,000. Whale-watching in and near these nursery lagoons is today a considerable tourist attraction.

The gray, like the humpback, has a face that is not exactly handsome. In fact it is downright ugly in both: pimply, warty, apparently misshapen and spotted with external parasites, such as whale-lice in clusters, and large barnacles to which may be attached living seaweed and goose-barnacles. These are acquired during the winter sojourn in semi-tropical seas. Nevertheless both whales are all grace as they swim and indulge in gyrations under water; their large bodies become supple and amazingly sinuous—as seen by the diver and recorded on film.

The gray (as it is spelt in the New World) is a medium-large species, up to 20 tonnes and 13.7m (45ft) long, a uniform grey varying in shade. The upper surface is irregularly dotted with warts or humps, some with bristles—and there are stiff hairs about the face; this 'hairiness', its apparent indifference to boats, and its slow speed when swimming, have earned it the reputation of being primitive—whatever that may mean. But they are not unintelligent. Females are larger than males. A strictly plankton and small-fish eater, having 150 pairs of thick baleen plates up to 45cm (18in) in length. The distensible throat carries two or four grooves 1.5m (5ft) long. Blubber acquired in summer feeding is up to 25cm (10in) thick.

From a close study of the population breeding in the bays of Lower California much is now known of its winter habits. The heartbeat has been measured at 27 per minute with the head above water and 9 per minute below. It is talkative, but will fall silent suddenly as if listening for some danger threat—such as the approach of its dreaded enemy, the killer whale. Like most cetaceans it appears to relay sound signals through layers of water of different density. The cow will reprimand an unruly calf with tail and flipper slaps. She is reported to decoy a boat away from the calf by swimming at the bow and distracting the attention of the inquisitive human visitor. Cousteau (1977) thinks it may be intelligent enough to pluck a harpoon from the body of a companion.

The gray is promiscuous; two males are recorded as serving one female on the same day. The males gather at the nursery lagoons ready to inseminate any oestrous female, but later they migrate in separate 'bachelor' pods along the coast on migration to and from the Bering Sea and Arctic Ocean.

Cousteau suggests that the cow will abandon her calf if it is weak or defective at or soon after birth. The child is otherwise jealously guarded by the mother who will attack and ram a boat, or other object, coming too close. The calf is large, up to 5m (16.5ft), and over 1 tonne in weight at birth. It grows rapidly in the shelter of the lagoons, much warmer than the ocean outside which is subject to the cool current from the north.

It is known that the gray whale migrates as rapidly by night as by day. Its migratory route from California along the coast and inner fiords of British Columbia and Alaska is haunted by killer whales feeding principally on salmon and large fish. But hungry orcas will at times attack

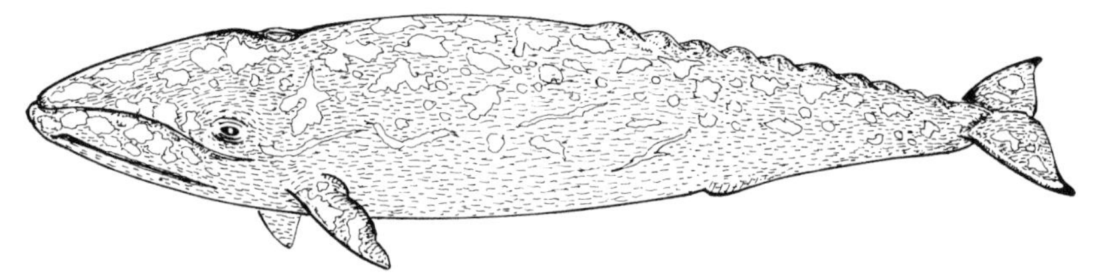

Californian gray whale Eschrichtius robustus (*formerly* Rhachianectes glaucus)

migrating whales and tear out lumps of flesh, leaving the victim—once appetite is satisfied—severely wounded or dead. In the summer of 1978 Bill Lawton, counting whales from the air off the north coast of Alaska, described to me how he watched and endeavoured to photograph a group of killer whales attacking a gray whale helpless at the surface, tearing the flesh and blubber in mouthfuls from the whale's sides and wrenching at the massive tongue—typical of behaviour observed when orcas are attracted by the copious blood of harpoon wounds to the bodies of whales alongside whaling ships.

The gray is said to be paralysed with fright on the approach of a killer whale. However, in nature generally it is the predator's intent to attack that matters, and so long as this is not evident to the prey the two whales will feed and swim peaceably within sight or hearing of each other (about one kilometre) as I and others have seen them do off the coast of Alaska.

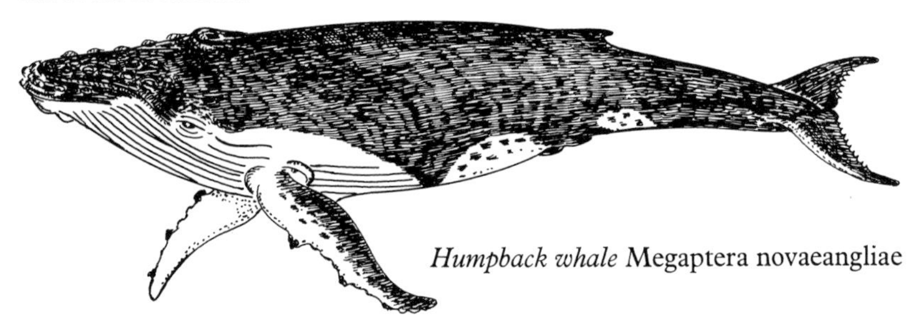

Humpback whale Megaptera novaeangliae

Humpback Whale *Megaptera novaeangliae*

Remarkable for its ungainly head, the narrow flattened upper jaws sunk into the great bulk of the lower jaws and bulging throat, like a knobless lid of a fat teapot. It also has the longest flippers of any whale—4.3 x 1.1m (14 x 3.6ft)—dark above and white beneath, with knobbly front edge. Despite its size, up to 18m (60ft) in length and over 40 tonnes weight, it makes frequent spectacular leaps clear of the surface, crashing sideways afterwards. Normally it swims near the surface, slowly at three to four knots, breathing once or twice a minute when the cup-shaped recess of the blow-holes briefly exhales the characteristic loud blow, a visible broad puff of watery breath some 3m (10ft) high. As the head submerges the rounded back with its short fin revolves into view (hence the name humpback) and the large tail with its narrow stock may or may not appear. Generally the tail does not appear, though it is a majestic sight when it does, adult flukes having a spread of 4.6m (15ft), often white with barnacles.

It is fairly easy to observe when engrossed in feeding in its summer quarters, close to the shore, as in Glacier Bay, Alaska, where for several summers Charles Jurasz has studied a group of some thirty adults, chiefly pregnant or lactating females, cruising within the beautiful environment of this national park. At his invitation I was privileged to join the Jurasz family in whale-watching there in 1978. Apparently aware that they are safe from human enemies here, these north Pacific humpbacks are tamer than the southern humpbacks I have watched in New Zealand waters. At times, by shutting off the engines, Jurasz was able to drift his boat within touching distance of a sleeping humpback. He has frequently swum with them briefly in these chill waters, and for longer in their warmer wintering quarters between the islands of the Hawaiian group.

Jurasz believes that to assist in rounding up its plankton and capelin food the humpback not only extends its long flippers but deliberately releases a chain of large air bubbles during its circling movement; the gurgling and reflections from these ascending balls of air cause the fish to mill together in a tight bunch. We could plainly see the great head of the whale rise in the centre as it completed the encirclement and the vast mouth opened; at the same time the throat pleats (of which there are two dozen extending as far as the navel) bulged and spread wide to assist the whirlpool of food and water being sucked into the enormous gullet. The jaws then clamped down, the great tongue rising to expel the water and retain the solids against the sieve of 300 pairs of baleen plates as the whale totally submerged again.

As a cloud of gulls and terns often circles close above the balling capelin and plankton when the whale rounds this food to the surface, it is not surprising to learn that sea-birds are occasionally found in the humpback's stomach—obviously they are sucked into that capacious maw by accident as they peck greedily at the disappearing food.

Like the shepherd who may ear-tag the individuals of his flock but also knows them by their individual appearance and habits, Jurasz, although he has occasionally planted a numbered identification dart in the upper part of some of his humpbacks, prefers to disturb them as little as possible, since he is able to recognise old and new summer residents in Glacier Bay by the varying shape of the dorsal-fin (a stub or inclined forward, backward or uneven outline), by the colour pattern, by the position of the barnacles on the skin of head, tail and flukes, as well as by size, habits and territory occupied. Jurasz finds that adults return year after year to the same territory, roughly about half a mile square, often close inshore in shallow water. When another whale trespasses the owner will blow more loudly, and if necessary will leap clear of the surface, making a prodigious splash to advertise its occupation.

Lob-tailing is another warning action, and if an invader persists in entering a territory, head-nodding and vocal sound-signals are further deterrents.

The voice of the humpback has been recorded both in its North Atlantic and North Pacific haunts. The two sub-species, although living so long isolated from each other, are found to have similar 'songs' (musical to the human ear, and now available as records or on tape), but do not show much interest when a north Atlantic humpback song is played back to a north Pacific humpback audience, and vice versa. Roger Payne has shown that this whale sings different songs in different years . . .'No other species is known to have such surprising behaviour. That they can memorise a new song each year constitutes evidence of a brain capable of storing and retrieving relatively large amounts of acoustic information.' Breathing is interspersed with song in a pattern which helps to identify the individual singer. Although songs both at Hawaii and Bermuda differ in each year, they sing some similar phrases in the two places. In fact they are 'musical' (in the human sense) and can improvise. Payne notes that 'evidence for 1977 shows that on the same date the Bermuda song extended at least as far as Silver Bank, 1,200km (750miles) south of Bermuda.'

It is thought by some students of the humpback's song that only the males sing at the winter quarters, to call attention to their presence and preparedness to mate. Contrary to a popular belief of whale-men in the last century that this and other baleen whales mate and keep together for life, the humpback is almost certainly as promiscuous as most other cetaceans.

Some 373 humpbacks were counted in Hawaiian waters in 1976; and 809 at Silver Bank, north of Hispaniola, West Indies in 1977. The warm seas of their wintering latitudes are necessary as nurseries for the new-born calf, which lacks insulating blubber at first; and here too the males assemble for the annual mating. The migration to high-latitude summer feeding grounds is in this order: females in early pregnancy, immature males and females, mature males, resting females, females in early lactation with new calves. First to return to the low-latitude wintering area are females at the end of lactation with weaned calf, then immatures of both sexes, mature males and resting females, females in advanced pregnancy.

From its preference for coastal waters, and its slow swimming, the humpback was very heavily overhunted, and was close to extinction early in the present century. Almost too late it was given complete protection by the IWC, but this has not been 100 per cent effective. It is still (1979) pelagically hunted off western Australian coasts, and a few from small boats by native fishermen of some Pacific and Atlantic Islands. Large pods are no longer seen, but it may be holding its own and per-

The tail of a southern right whale (Jen & Des Bartlett/Bruce Coleman)

haps slowly increasing. The southern subspecies feeding in sub-Antarctic seas, migrates north past both coasts of Australia, New Zealand and South Africa to winter south of the equator.

TOOTHED WHALES (The Dolphins)

Sperm Whale or Cachalot *Physeter catodon*

Largest of the toothed whales or dolphins, in which the male is generally larger than the female. The adult male cachalot is more than one-third as large again at 20m (66ft), and three times as heavy (up to 58 tonnes) as his mate. This dolphin is unique in its deep-water hunting technique, and in having a head one-third as large and heavy as its body. There are no subspecies; it ranges all oceans, travelling in large pods but some of the old males become solitary and feed during summer in high latitudes. Adult females are less migratory, remaining with their calves and immatures within latitudes where the water temperature is fairly warm (20° to 30°C).

Squid and octopus are its favourite food; its feeding habits and curious anatomy have been described in earlier chapters. Wherever squid are numerous the cachalot will not be far away, from equator to the polar ice edge. It is still the most numerous and widely distributed of the great whales.

A family group usually consists of up to a dozen pregnant or lactating females and their young, including new and older calves (up to five or six years old), and a master bull who dominates the herd and will drive away rival males. This polygamous hierarchy results in the formation of pods of exiled bulls which are regularly found far from the maternal or harem groups. It is believed that these non-breeding male 'clubs' may persist as permanent pods, indulging occasionally in homosexual behaviour, and at intervals fully mature members making forays towards family pods to test the defence put up by the resident master bull. It is difficult to explain the presence of large old-looking bulls haunting waters close to zero temperatures in polar seas hundreds of miles away from the family groups except as spent individuals content to browse in the squid-rich seas of high latitudes.

Observation of family pods suggests that each female closely guards its child up to two or three years old, and will repel the sexual advances upon their young daughters by older but still immature males born in or attached to the same pod. The cow is a good disciplinarian but also a fierce fighter i n protecting her dependent child, as the whale-hunter has frequently described. It was unsafe to harpoon the calf first, for its cries of distress would frequently result in the mother attacking and smashing the whaler's small boat with her snout, jaws or tail, or all three. Better to harpoon and kill the mother first, then the bewildered suckling would

A gray whale's tail (Kenneth W. Fink/Ardea Photographics)
A humpback whale breaching (Charles Jurasz)

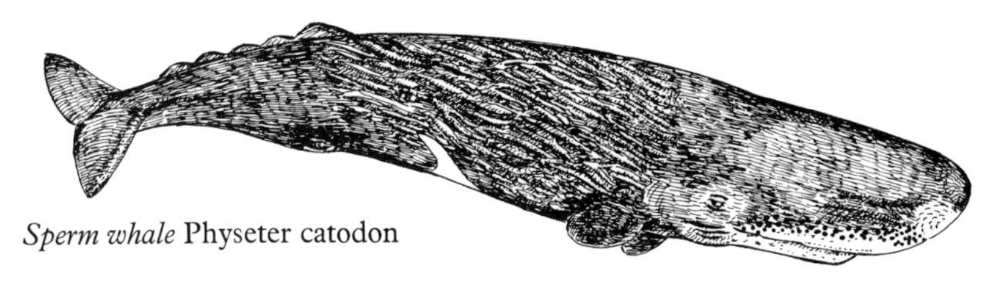

Sperm whale Physeter catodon

swim to her body, convenient for the slaughter.

Albino individuals frequently occur, more than in other whales; Moby Dick of Herman Melville's stirring novel was an albino male. One of the first and perhaps most reliable descriptions of sperm-whale hunting is that by Fredrick Debell Bennett (1840) in his narrative of a whaling voyage round the world, 1833–36. He was the first to note distant underwater communication between sperm whales: 'It is confirmed that upon a cachalot being struck from a boat, others, many miles distant from the spot, will almost instantaneously express by their actions an apparent consciousness of what has occurred, or at least of some untoward event, and either make off in alarm or come down to the assistance of their injured companion.'

During mating activity in the winter months, when the family pods are in sub-tropical waters, there is much playful leaping clear and crashing back upon the surface; but this behaviour was also considered by the old whalers to be a signal of danger upon which the pod would dive and disappear.

Ray Gambell, from statistics of teeth-growth rings and examination of organs at whaling stations in South Africa, has estimated the gestation period to be 14¾ months. The new-born calf (occasionally there are twins) weighs 1,000kg (2,200lb) and measures 5m (16ft) in length. It is suckled for longer than any cetacean so far recorded—25 months—after which the cow enjoys a rest period of 8 or 9 months before conceiving again. Sexual maturity (ovulation) in the female is at 7 years when she measures 8.8m (30ft); she continues to grow until physical maturity is reached in another 20 years at 12m (40ft). Spermatogenesis may occur earlier in the male cachalot, but he does not achieve dominance as harem master until he is about 26 years old and 12.5m (41ft). He continues to grow until he reaches physical maturity at 40 years when his maximum weight is nearly 60 tonnes and length 20m (66ft). This long period of growth and the four-year interval between calvings suggests that the cachalot may live at least as long as a human.

This black or dark-brown whale exhibits a white gape, which it is believed may attract its prey towards it as the cachalot quarters the almost lightless depths to which it descends. The blow-hole is well forward and S-shaped—55cm (22in) long, placed to the left of the tip of the enormous snout or head. This gives the blast, varying between 5-15m (16-50ft), a forward rake of 45°—a distinctive characteristic since other whale-spouts are perpendicular. When sounding the tail is invariably lobbed into the air; the dive is almost vertical. The dorsal-fin is a mere flattened series of knobs forming an uneven ridge towards the narrow tail-stock. The blubber is up to 30cm (12in) thick.

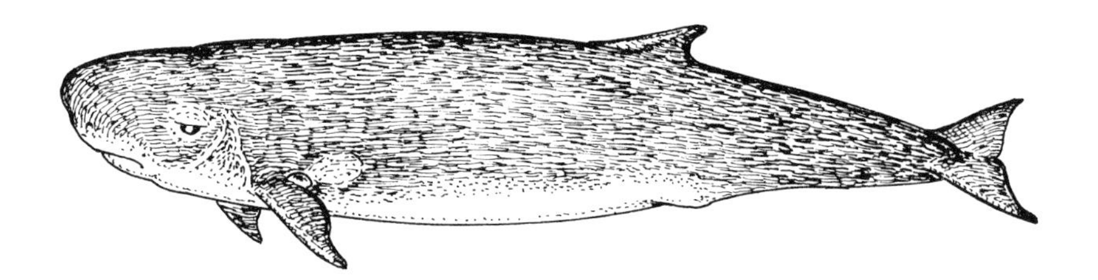

Pygmy sperm whale Kogia breviceps

Pygmy Sperm Whale *Kogia breviceps*
Until recently this diminutive relation of the cachalot, only 4m (13ft) long and up to 680kg (1,500lb) in weight, was considered very rare. Yet it is recorded in all temperate to tropical oceans, chiefly through single strandings (20 in New Zealand) which seem to have been caused by distress from internal parasites (Robson, 1976). Virtually nothing is known of its migration or breeding. Like its giant relative it has sharp recurved teeth in the lower jaw only; these fit into recesses in the upper jaw. The blow-hole, however, is situated nearer the top of the large head, above the eye, and somewhat to the left of centre. It seems to be a deep-water species, and probably feeds on squid.

Narwhal *Monodon monoceros*
The original unicorn—the male is the only naturally one-horned mammal in the world. The whole of the upper parts are heavily spotted or marbled with brown against a paler background; a low ridge 50cm (20in) long replaces the dorsal-fin. Length (excluding the tusk) up to 5.5m (18ft) and weight rarely up to one tonne. Its present distribution is limited to the eastern Canadian arctic and ice-laden seas between Labrador, Greenland, Spitzbergen and the White Sea. An aerial count of its numbers in the east Canadian Arctic in 1978 produced a maximum estimate of 20,000. In the whole of its range the total population may be much less than 30,000.

In embryos two pairs of teeth develop; the posterior ones disappear before birth, leaving one in each half of the upper jaw. In the male one tooth (rarely both), usually the left, grows forward in the same axis with the body to reach a maximum length of 2.7m (9ft). It is hollow and twisted in a tight spiral to the left. The ivory tusk has a considerable value today as a curio, and for carving in Eskimo art; and this demand has led to depletion of numbers which gives concern for its survival. Old females sometimes develop a very short tusk.

As to the significance of the tusk, it does not seem to be an adequate

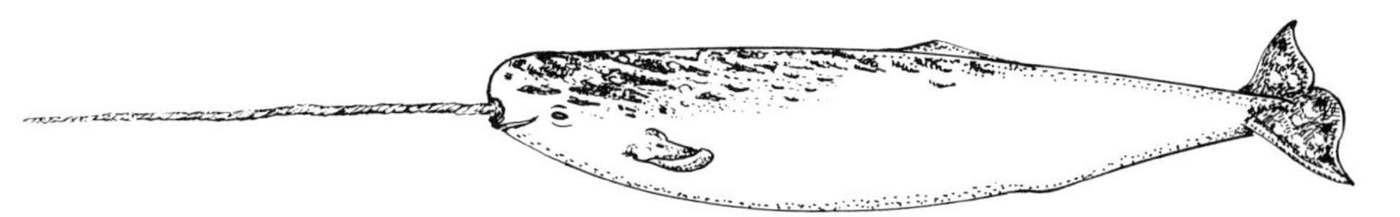

Narwhal Monodon monoceros *male*

Top view of male narwhal's skull

Narwhal Monodon monoceros *female*

weapon for defence or food-gathering. In fact it is difficult to understand how the male can gather its food, said to be quite large cod and squid, with its long tusk protruding above the comparatively small mouth. It would also seem to be an embarrassment in winter, when this dolphin rests and breathes at a breathing hole in the ice, since, although the tusk may be used to enlarge the hole (I can find no observation in the literature supporting this possibility), its appearance poked above the ice must betray the narwhal's presence to its chief predators, polar bear and man.

Alan Best, former director of the Vancouver Zoo, collecting narwhal for the aquarium, has seen a pod of male narwhal at rest; they formed a circle in shallow water, facing outwards, and some rested the long tusk above and upon its neighbour's. If this is a common habit of this sociable dolphin, perhaps the tusk is largely a status symbol of sexual significance, used, like the large comb and wattles of gallinaceous cocks, to impress and dominate, rather than in fighting. There are one or two reports (Bryuns, 1971) of 'fights to the death between this species and the walrus for breathing-holes in the pack-ice.' The tusk is considered to have aphrodisiacal properties; a Baffin Island Eskimo sculptor told me he was able to sell at a good price the shavings and dust accumulated on his work-bench during his carving of narwhal tusks for the tourist trade. They were bought by an agent in Hong Kong trading in commodities (including rhino horn, and bone from the penis of walrus and seal) alleged to stimulate virility in man.

Incidentally the narwhal netted by Alan Best and conveyed by air to Vancouver did not survive long. This species is naturally heavily infested with lungworm, and despite antibiotic treatment has so far never been kept successfully in captivity after removal from the chill waters of its home.

Diving constantly under the pack-ice the narwhal communicates with a whistling call over long distances under the water. Pairing takes place in April and the calf is born after 14.5 months gestation during July and August. It has a 2.5cm (1in) coat of blubber and swims close to its dam; the pair call softly to each other as they follow the melting pack-ice northwards from August onwards. Lactation probably lasts at least 18 months. Much remains to be recorded of the breeding biology of this unique, strictly arctic dolphin, so rarely observed in winter. Its range appears to be diminishing; it is being increasingly and wantonly killed by gun-happy Eskimo and other hunters (some are irresponsible youths) who no longer trouble to utilise the carcase fully; in any case about 50 per cent of the wounded or dying sink; some 500–600 are killed annually and brought to land. The head is cut off, so that the teeth and tusk ivory can be extracted later; and some of the skin may be removed. Narwhal skin is considered a delicacy notably rich in vitamins and has a sweet taste; it is eaten raw by Eskimo people.

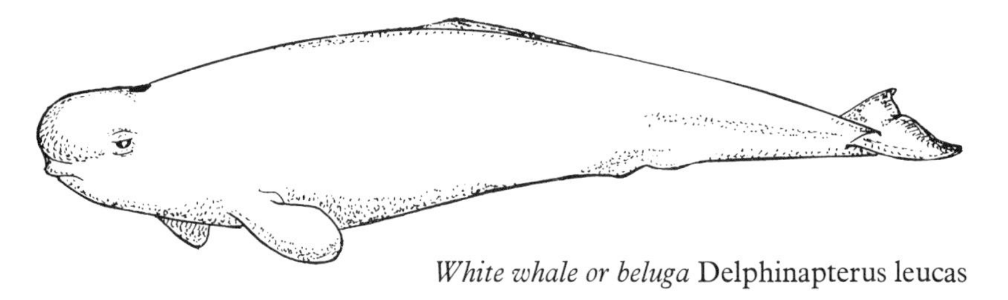

White whale or beluga Delphinapterus leucas

Beluga or White Whale *Delphinapterus leucas*

Probably there are about 30,000 of this beautiful pure white whale inhabiting all coasts of the Arctic Ocean along the edge and in the leads of the pack-ice. Their bodies are inconspicuous against snow and ice, unless they move among the ice floes, or in open water, as seen from the air, or the high prow of a ship. They are gregarious and noisy—whalemen dubbed them sea-canaries. I have heard their cat-like mewing in the Canadian Arctic as well as in the tanks at Vancouver Aquarium, where they have been taught simple tricks (such as spitting water at their trainer) and their graceful flowing bodies are delightful to watch through windows in the side of the tank.

Family parties will accompany bowhead and other baleen whales as these feed in the plankton-rich Arctic Sea; there must be practical reasons for this association: both watchful for enemies, both rounding up krill and associated fish for mutual benefit. The beluga feeds on prawns and medium-sized fish. In length and weight it is approximately the same as the narwhal, but the beluga makes a longer winter migration, south to the Aleutian Islands in the Pacific and to the Gulf of St Lawrence and Norway on the Atlantic side. Immature individuals occasionally swim far up rivers and have been reported over 1,100km (700 miles) up the Yukon. One lived for a while on the Rhine, beyond Holland and far into Germany, in the summer of 1966. Belugas which enter tidal rivers have been caught by placing steel rods upright across the waterway. Although these may not be so close together as to prevent the whales squeezing through, the echoes of their sonar clicks received back from the obstacle and analysed in the beluga brain warn them that the barrier is impassable. Hemmed in by men in boats upstream who drive them towards the barrier, and sentinels in anchored boats on the downstream side, both beating tin cans, the beluga are confused and trapped on the falling tide. They are gentle creatures, with small teeth, and affectionate towards a kind keeper. The first to be born in captivity lived for six months at Vancouver.

The calf is born in the summer (July–August) after a gestation of one year. It is dark grey-brown at first but assumes a delicate cream-white, sometimes flushed with shell-pink, but the time it is three years old. Lactation lasts for at least eight months.

The soft bulbous melon forward of the blow-hole indicates a good acoustic 'eye' for sound scanning in the twilight under thick ice, and the location of breathing holes which these dolphins maintain when the bays and fiords of their Arctic home are frozen. Sometimes the beluga is trapped by heavy ice blocking the shallow entrance to a small bay where it has been feeding. Under these conditions it may run short of food, and weaken, become less alert to predators lurking close to its breeding hole. The polar bear is capable of crushing the beluga's head with a single blow of its huge clawed paw. According to a Soviet observer at Novaya Zembla, one of these bears killed thirteen belugas, one after another in this manner as each member of the pod reared its head to breathe, and then dragged them out onto the ice. Soviet observers also say that belugas are attacked by the orca 'when they panic'.

As the bull beluga is larger than the cow, it is likely that a master male dominates each pod and maintains a harem for as long as he is able to drive away rival males.

Orca or Killer Whale *Orcinus orca*

This handsome piebald whale is one of the largest of the typically dolphin-shaped toothed cetaceans. At maturity males reach 9m (30ft) in length and weigh up to 8 tonnes; females about one-third less. It is immediately identified at sea by the tallest dorsal-fin of any cetacean, up to 1.8m (6ft) in the male, and often held high above the surface for much

Killer whale Orcinus orca

longer than most dolphins; the female back-fin is half as tall and usually more recurved.

The curious distribution of white zones on this black whale seems to have significance as a guide to identity, especially perhaps between cow and calf. There is a white stripe along the keel of the belly, covering the genital-nipple region; the white throat, flanks and white oval above the eye may help to synchronise swimming side by side. A group of orcas will sometimes stand on their tails (so to speak) and stare at the human visitor, with head held vertically. This is called spying, and is generally followed by further concerted action: the pod may retire, or advance to investigate further. Among pack-ice, having located in this way seal or penguin resting on the floes, killer whales have been known to dive, and dash their snouts against and capsize the ice-rafts in order to seize and devour the occupants. The case is often quoted of how the Scott Polar Expedition cameraman Ponting was nearly thrown into the sea by floe-ramming orcas. Whether they would have devoured him on finding him wrapped in inedible clothing and boots instead of tasty blubber was fortunately not put to the test. But it is curious that there appears to be no eyewitness record of this carnivorous whale ever hunting down and eating a man, but several instances of a man falling overboard into water infested with orcas without being harmed by them.

There are few reports of action taken by orcas disturbed by the behaviour of man or other animals. Barking dogs seem to annoy them and one such was killed by a captive orca which seized it at the edge of the aquarium tank. There is the tale of two lumberjacks skidding logs into the water in British Columbia in 1956. One man deliberately released a log so that it struck and bruised one of a pod of orcas cruising close to the shore. The whales swam away, but that same night, as the lumberjacks were rowing back to camp, the pod reappeared and capsized the boat. The man who had released the offending log was never seen again; his companion survived unharmed to tell this story. Those who, like Frank Robson, believe in telepathy between man and cetacean, may be willing to believe this circumstantial tale.

In captivity, the orca has proved just as intelligent and gentle towards man as the bottle-nosed dolphin. It is as playful and good at athletic feats. At Vancouver Aquarium I watched a male, at the bidding of its trainer, make a prodigious leap 25ft into the air, turning over and re-entering the water head first in quite a small tank—a considerable feat of timing skill and courage in this massive bull which weighs probably 6 tonnes. Afterwards it swam on its back and clapped its flippers together amusingly, as if applauding itself.

As a result of this docility and learning skill, and to satisfy the demand from aquaria, a considerable live-capture orca fishery sprang up where this dolphin is common, in the quiet sounds from Seattle north through British Columbia to southern Alaska. Between 1962 and 1973, 263 orcas were netted, of which 50 were sent to oceanaria; over this period the revenue to the netters was estimated to be one million dollars; this hunting led to a decrease in the number frequenting these sounds. Permits are now necessary to take orcas, which 'reflects an increasingly cautious policy of the two governments (of Canada and the United States) with respect to cropping until research on population abundance can establish harvest limits' (Big & Wolman, 1975).

Demonstrating the genuine affection which the orca can display towards man, one individual has allowed its mouth to be held open while the trainer has thrust arm or head within range of being cut off by one snap of these formidable sharp pointed teeth—twenty-four in the upper jaw interlocking with the same number in the lower jaw! Occasionally the captive orca has tried to detain its human companion by gripping a limb, but so gently that the teeth have not broken the skin. Perhaps emboldened by this evidence of harmlessness towards man, a number of persons have of recent years paddled a light canoe or kayak among the pods of orcas which spend the summer feeding within the sounds of British Columbia. New Zealand-born Paul Spong, a director of Project Jonah, Canada, has stated that he swam with wild orcas in perfect safety; adding that 'we often play music to the whales, for we feel their interest in and perhaps their appreciation of it. Sometimes, particularly on a still night, a pod or part of a pod, or perhaps just a single whale, will hover off-shore for an hour or more tuning into the music. Sometimes they seem to join the celebration with the chorus of their voices and the dance of their bodies, visible from the bubbling phosphorescent wakes they leave behind.' The killer whales of these sounds have since been filmed several times; Spong believes that there is 'unquestioned communicative exchange between free orcas and humans' (Spong, quoted from McIntyre, 1974).

Yet the orca has for centuries had the reputation of being a ruthless killer, hunting in packs and rounding up its prey, including man-sized dolphins and seals, porpoises, sharks, salmon and other large fish, as well as baleen whales ten times the bulk of an orca. During the encircling movement the orcas grab any individuals attempting to escape, often tossing them high into the air with savage playfulness, causing wounding, further disarray and panic. Each orca in turn goes in among the milling victims, to kill and maim, while the other orcas keep the net closed. The pack then feeds more peacefully on the disabled and dying prey, allowing those which are unharmed to escape.

Like a cat with a mouse, the orca will play with its captive, disabling it, but, if not hungry, leaving it when it ceases to struggle to escape or is dead. Virginia Jurasz tells me: 'One of the saddest things was to watch a pod of four killer whales play with a small seal. The seal swam to the *Ginjur* and at first tried to climb into the Zodiac dinghy. When it was unsuccessful the seal attempted to flatten itself against the hull of the *Ginjur*. It looked as if it was already injured. The female killer whale would sweep up, grab and pull it under, only to release it again. It was almost like watching a cat play with a mouse. Everyone on board was touched by the seal's sad brown eyes, the demands of nature, and man's lack of knowledge of how to deal with those demands.'

Two oft-quoted records of the orca's appetite have never ceased to cause wonder. According to Slijper (1962) 32 seals were found in one orca's stomach, and 13 dolphins and 14 seals in another. However, investigation of the source proves them to have been fragments of bones digested over a long period.

Theoretically, with no other predator than man, orcas should be abundant. Yet over the oceans as a whole they are not at all common, except where, as off British Columbia coasts, they gather seasonally to feed on migratory salmon and other large fish. It is evident that their numbers remain in proportion to their food supply; and it is probable that their predation of warm-blooded penguins, seals, dolphins and large whales is limited, and even healthy, by the removal of weaker individuals less fitted to survive, as wolves have a tonic effect on the health of caribou. Like the lions and tigers of the land, without man to control their numbers these lords of the ocean are ultimately controlled by those natural consequences of overpopulation: contagious disease, proliferation of internal parasites, and malnutrition, inhibiting breeding success.

Charles Jurasz has described to me the scavenging role of the orca off southern Alaska, and how healthy humpbacks successfully 'stood off' their lethal enemies. He was puzzled by what at first seemed to be a pair of humpbacks chasing five orcas ahead of them, and now and then rolling over violently and gasping. Soon he discovered that four other orcas were attacking from the rear, darting at the vulnerable belly from below, 'punching and feigning', but not biting because at the crucial moment the humpback inverted swiftly, after gulping a breath, causing the sea to boil around its body in a foam-edged swirl. Other humpbacks moved up to form a tight formation which made the pod of nine orcas slacken their efforts, although 'the waxing and waning of the humpback pods, the attacking or testing approaches of the orcas, dispersals and regroupings continued for three hours—with the *Ginjur* at the ringside—for the staging area's mobility drew our boat to centrestage.'

Perfect health and the assistance of other healthy members of a pod enabled the individual to resist the at first unrelenting testing of the body's weakest point by the combined force of nine orcas working to a concerted plan. A lone humpback in poor health would have been torn to pieces by such a formidable killer gang. As it was the attacked individual in this instance had a hard time of it at first, twisting its great body to cause a swirling which deflected the punching thrusts, gasping and bellowing and using energy at a vast rate. What a conversation must have been going on! The orcas talking perhaps, exchanging signals between the decoy group upfront and the rearguard attackers. But the calls for help were answered; the other humpbacks moved in, and the orcas eventually swam away—demonstrating once again the importance for survival of gregariousness, of coming together in defence, which is characteristic of all the cetaceans.

Although the orca is found in all oceans, the extent of its migration is unknown. Nor has its breeding biology been fully studied. The calf is born in late summer, after a gestation period of about 15 months; it is 2m (6.5ft) long, the white having at first a yellowish tinge, possibly from the amniotic fluid of the birth-bag. It remains close to its mother for more than a year. Pods counted near Vancouver have consisted of a large male and usually two cows with their calves, both new and older ones. Because they often swim abreast or in line formation and breach simultaneously they have been called respiratory units by Spong who finds that the same adults, identified by fin-shape and angle and distribution of the white pattern, return year after year. Senior bulls will visit different pods and are the only class seen travelling, or at rest, quite alone.

Young orcas are playful, leaping out of the water as the pod moves quietly through the calm sounds. Cousteau has recorded an underwater ballet of joyous gyrations performed by wild orcas: 'Their lives indeed seem to be almost ideal and, from their known intelligence and apparent interest in music, perceptive of ours. They glide past and turn, watching us unafraid and curious, friendly so long as we are friendly. We stare in wonder at their grace and beauty of form as they move close to our boat, leisurely feeding, playing, dozing, teaching their young how to live, and breathe in unison.'

They appear to have no care or enemy save man and old age. They are the lordly ones, the blithe, beautiful, kind and terrible among the kings of the ocean.

The Beaked Whales *Ziphiidae*

A family of medium to small whales, dark-coloured with tapering beaks, and (except for the Tasman whale) with only two teeth, which in the male are used in fighting and dominance, like the spurs of a gamecock.

Most of these whales, including females and immatures, carry healed scars in the form of white lines on the upper part of the body. In the females the teeth are shorter or do not erupt. The dorsal-fin is very short.

Not much is known about their life history. When harpooned they dive almost vertically; in general it is considered that the beaked whales are oceanic and deep divers, feeding much on squid and soft fish. Having no teeth effective for crushing they must gulp down the food whole. Some are known only from rare strandings.

Tasman Beaked Whale *Tasmacetus shepherdi*
Known from six washed ashore in New Zealand. Five were males 6.1–9.1m (20–30ft) long. One female 5.5m (18ft) long. Estimated weight of a large male is over 3 tonnes—a useful size for the whalers of the sub-Antarctic, which it probably frequents, but there are no records of it being harpooned. It has numerous small teeth in both jaws, and two large 'fighting' fangs near the tip of the lower jaw which place it in the beaked whale family.

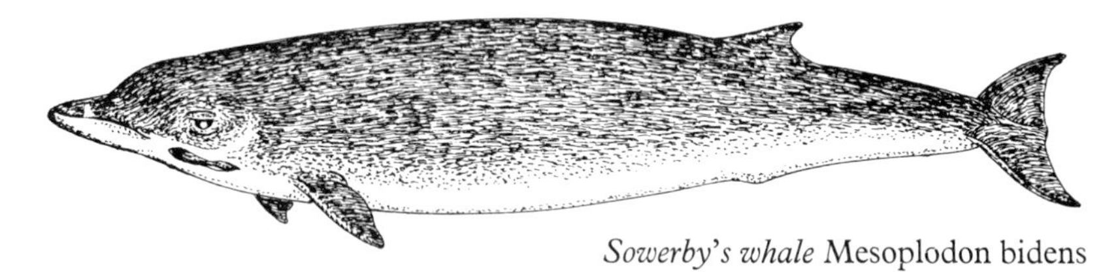

Sowerby's whale Mesoplodon bidens

Sowerby's Whale *Mesoplodon bidens*
About 5m (16.5ft) long. Found in the Gulf Stream area of the North Atlantic, along with the next two species. Rarely or never reliably reported at sea, but known from strandings on Gulf Stream coasts.

True's Beaked Whale *Mesoploden mirus*
Also known as the Antillean Beaked Whale, and resembles the last, but up to 5.2m (17ft) and 400kg (3,100lb). Almost toothless; two small teeth erupt late in life in the male, and not at all in the female. Measurements were obtained from 14 stranded individuals on North Atlantic coasts. Mysteriously, one was stranded on the coast of South Africa in 1960.

Gulf Stream or Antillean Beaked Whale *Mesoplodon europaeus*
Larger still, up to 6.7m (22ft) and 2,700kg (5,950lb). Known only from a dozen strandings including a mother with a new-born calf in Jamaica in 1953.

Scamperdown Whale *Mesopoldon grayi*
Like the last, the female has no teeth. This small dark cetacean, up to 4m (13ft) long and 770kg (1,700lb), is an inhabitant of temperate to sub-Antarctic waters. Twenty at least have been stranded on New

Zealand coasts, off which Bruyns records it as jumping freely, showing a typically scarred and spotted skin. One has stranded in South Africa.

Hector's Beaked Whale *Mesoplodon hectori*
Only known from skulls found in Tasmania and New Zealand.

Longman's Beaked Whale *Mesoplodon pacificus*
Even less known, only from a skull in Queensland and another skull described from Mogadiscio, Somalia.

Sabre-toothed Whale *Mesoplodon stejnegeri*
So-called from the pair of teeth which are visible as they project from the centre of the lower jaw. Up to 4.9m (16ft) and 1,270kg (2,800lb), known only from a dozen strandings, Japan and Oregon coasts north to the Bering Sea. The skin is well scratched and spotted from the two large fangs of the male; the female also has two visible smaller teeth.

Arch-beaked Whale *Mesoplodon carlhubbsi*
Known only from five strandings on North Pacific coasts. *M. bowdoini* is known from half a dozen individuals washed up on New Zealand and Australian shores.

Ginkgo-toothed or Japanese Whale *Mesoplodon ginkgodens*
Known only from three examples caught separately on lines in the Japan Sea; these were 5.2m (17ft) long and 1,450kg (3,200lb). So called because the two large teeth of the male are compressed sideways and shaped like the leaf of the ginkgo tree. The female has smaller teeth. The body is covered with healed scars and also with round white spots.

Strap-toothed Whale *Mesoplodon layardii*
A southern ocean whale known from strandings chiefly in New Zealand and largely of females. Males have a long pointed pair of fangs, females much smaller ones. Length to 4.9m (16ft) and weight estimated 1,270kg (2,800lb).

Dense-beaked Whale *Mesoplodon densirostris*
Almost identical with the last, both dull-coloured and well marked with healed scars. The fighting teeth project nearer the hinge of the jaws. Known from under twenty stranded specimens all on tropical coasts. Length 4.6m (15ft) and weight up to 1,100kg (2,400lb).

Cuvier's Whale *Ziphius cavirostris*
Closely related to the beaked Mesoplodon species above; the same large pair of male fighting teeth in the lower jaw inflicting long scars on the body; these teeth do not erupt in the female. This is a large dark whale, males to 6.7m (22ft) and females slightly larger to 7m (23ft). Weighs up

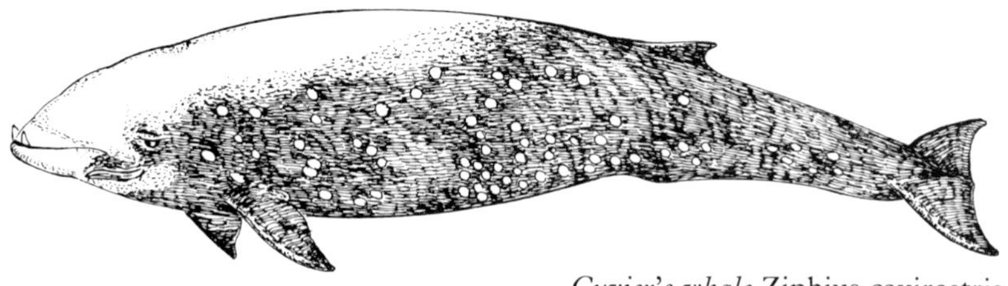

Cuvier's whale Ziphius cavirostris

to 6 tonnes. A temperate-water species, with many strandings in New Zealand, some in Japan, South America and Europe. When observed from ships they have been swimming lazily, jumping rarely.

Giant Bottle-nosed Whales *Berardius*
Two species, the southern *B. arnouxi*, 9.8m (32ft) and 8 tonnes, in Antarctic waters near New Zealand and South America. The northern *B. bairdi* is found in the North Pacific—Japan, Okhotsk Sea and Aleutians to Oregon. The latter is heavier, up to 12m (39.5ft) and 11 tonnes, a substantial prize for the Japanese whalers who used to take about 300 annually and still may do so, although they are scarcer today. Except for their size they closely resemble the Mesoplodon beaked whales, dull-coloured with numerous long white scars from the use of the two (only) fighting teeth of the males. However in all these two-toothed whales the wounds inflicted seem to be quite superficial, as if in using them as a weapon of aggression, dominance could be achieved by a gentle scratching of the teeth. Savage fighting would be wasteful, and destructive.

Bottle-nosed Whales *Hyperoodon*
The diversity of form in the whale is astounding. In this genus there are two medium large whales with typical Mesoplodon beaked jaws, the males have one pair of fighting teeth at the tip of the lower jaw, which do not project beyond the upper lip. In the female the teeth are very small and may not erupt. The main feature is the enormous forehead or melon which overhangs the beak, and is clearly a special aid to deep-diving sonar and location of prey.

Hyperoodon ampullatus
Ranges the temperate to arctic North Atlantic, males to 10m (33ft) and 8 tonnes, females to 8.2m (27ft). From its size this brown-black whale was eagerly hunted wherever it could be found by whaling-ships. It has the loud blast of a deep diver and when harpooned goes down almost vertically. The calf is born in early summer, when pairing also takes place. Gestation period one year; the female probably does not calve more often than once in three years. Up to 3,000 were taken annually by whalers under sail and steam; in the present century it has become rare. The huge head is a reservoir of spermaceti. On 20 August 1958 a

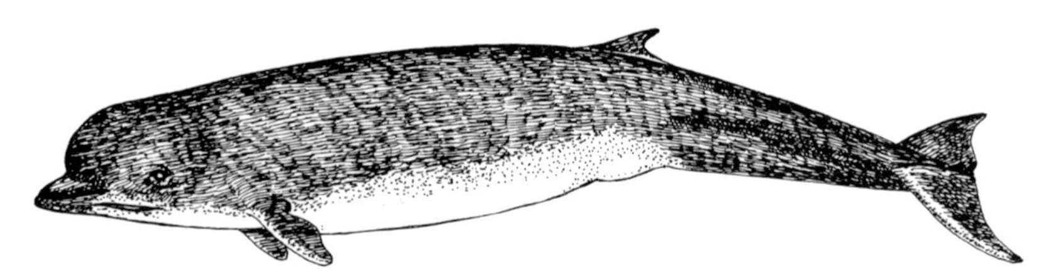

Bottle-nosed whale Hyperoodon ampullatus

live example 7.3m (24ft) long was found trapped inside the hull of a small coasting vessel which had sunk off Flushing, Holland; it was killed by the knife of a diver.

Hyperoodon planifrons

The smaller southern representative, ranging Antarctic and temperate latitudes. Males to 7.6m (25ft), females to 6.7m (22ft). It is seldom seen. Since 1862 only a few strandings have occurred, in Australia, New Zealand, South Africa and South America. The dorsal-fin is short, well to the rear of the back.

The little whales: Dolphins and porpoises

The dolphins lord it greatly among the herds of the sea, pluming themselves eminently on their valiance and beauty and swift speed in the water; for like an arrow they fly through the sea, and fiery and keen is the light they flash from their eyes.

Oppian

Happy, sociable dolphins, about which we now know so much from watching them in modern aquaria and oceanaria where, under as ideal conditions as captivity permits, some may live for several years and a few species breed self-perpetuatingly. It is now possible, with the addition of many observations of dolphins at sea, to reconstruct a model life history for a number of these smaller, toothed whales from records of captive and controlled individuals: the bottle-nosed, common and spinner dolphins, the pilot, false killer and orca whales, all of which have done well in captivity on easily procured and cold-stored fish. The pleasure and stimulus given to the thousands of people who visit oceanaria must help to make these visitors conservation-minded.

Because of their size and requirements of large tonnages of fresh plankton and crustacean krill, it is much more difficult to keep large baleen whales in captivity, and so far no success has been achieved. Their life histories have been described as far as they are known in the previous chapter. There, too, it has been convenient to include those interesting toothed giants the cachalot, the orca and other medium large dolphins, and the unusual narwhal and beluga.

It remains to describe the smaller, toothed whales, most of which are so alike in their several families that even the expert observer is unable to be sure of separating and identifying related species from the brief glimpse one usually gets of dolphins at sea—flash of snout or fin, or a distant leap. Most of them are rarely seen, and some are known only from one or two individuals washed ashore. As far as is known all of them give birth to the calf in the warmer latitudes of their range: the new-born is lean and needs warmth.

The chief features of the life histories of the cetaceans so far studied at sea, in captivity and by dissection after death, which may serve as a guide and informed guess as to the lives and habits of the rare and unstudied, have been briefly sketched earlier (see Chapter 2, under Reproduction).

DOLPHINS

Bottle-nosed Dolphin *Tursiops truncatus*

Much of our present knowledge of the behaviour of dolphins in captivity is derived from this species, which lives in temperate to tropical oceans, from 15° Celsius upwards, and comprising two not very well defined races. *T. t. truncatus* ranges the Atlantic and Indian Oceans including the Mediterranean, Persian Gulf and Red Sea; *T. t. gilli*, darker, almost black, inhabits the eastern Pacific within these latitudes. Both are the same size, up to 3.7m (12ft) in length, and to 370kg (800lb) in weight. They can be confused with the common dolphins *Delphinus*, having the same beaky 'nose', but grow much larger. Both enjoy playing around ships and riding in the pressure wave at the bow, easily keeping up with a ship travelling at 20 knots, but after a while this speed tires them; they will remain longer with a boat travelling at half that speed.

The first self-perpetuating colony of *Tursiops* was established at Marine Studios, Florida. Here the birth was witnessed by man, and the tender mother-care, and the assistance by a 'midwife' companion to lift the new-born to breathe at the surface, as already described. The mother normally exhibits a zealous and jealous care, keeping the calf constantly at her side, touching or cradling it with her flippers, guiding it away from disturbances and inquisitive visitors, even from the helpful female companion (who may be her own daughter). In this way the young calf is trained to obey mother, and meekly does so at first; but after a few weeks it causes maternal anxiety by exploratory forays and attempts to play with more distant objects. Mother can be boring at times. Usually it returns at her call to the safety of the play-pen of her flippers.

If it is slow to return she may punish it; in one instance a wandering calf was forced to the floor of the tank by its mother, who held it down until she herself needed to rise and breathe at the surface. In another apparently punitive action a mother slid upside-down under her calf and holding it in her flippers exposed it fully to the air for a whole minute. The calf tried in vain to wriggle free. Such treatment was followed by the calf swimming obediently close to the mother for the rest of the period of observation.

In the wild this close body contact between mother and child can be seen in this and common dolphins sporting in the bow-wave of a boat travelling at speed. In both species I have watched the half-grown calf keeping close under the maternal fin, its own small fin touching her body frequently as, with two or three other adults, the pair surged forward in the pressure wave, their sleek bodies superbly synchronised in swaying swimming movements.

As studied in captivity this close attachment lasts until the mother becomes pregnant again, or her milk ceases to flow. But much earlier, within weeks of birth, while the mother's vagina is still enlarged from parturition and she is in a semi-oestrous condition (as in many mammals,

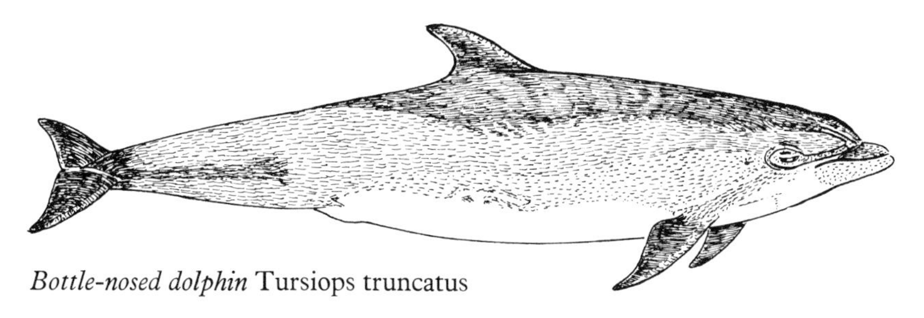

Bottle-nosed dolphin Tursiops truncatus

notably seals) she may permit, even assist, a male calf to achieve intromission during its playful activities around her body. Involuntary erection of the baby penis occurs frequently; and evidently through their common pre-natal blood circulation he shares the active hormonal condition of the maternal body for some time after being born. It is a natural act, and doubtless the instinctive affection between mother and child is strengthened by this intimate re-entry of a part of the infant body within the sensitive genital region from which it so lately emerged.

This and further incestuous behaviour is frequent in dolphins of all ages in captivity, and much of it must arise from the sheer boredom of imprisonment in this intelligent, inquisitive animal. In the wild, although dolphin mother and child are seen to play with each other during resting periods at sea, their normal swift pursuit of fish and their strenuous voyages must reduce the time, opportunity and inclination for idle erotic and masturbatory behaviour.

The dolphin begins to eat a little fish, after preliminary mouthings and rejection, at about four months, and is normally weaned from milk altogether between a year and eighteen months. Advanced pregnancy usually induces in the mother an indifference to the yearling calf, which has already learned to play games and form attachments which build up in these intensely sociable dolphins. Amusingly like older daughters in a human family, adolescent females will adopt (sometimes try to steal) suckling infants and older calves and herd them away from the dangers of violent activity elsewhere in the tank, such as mating pursuits, or leaping and splashing, or when the mother is being fed by the trainer and her attention is momentarily withdrawn from her child. The nursemaid or maternal instinct, like the human girl's enjoyment of playing with dolls, develops early in the female dolphin.

As in most sociable mammals there is an established but periodically changing order of dominance, based on the rule that might and courage are right. The mother will defend her child with all the courage which gives her the strength to attack an adversary, and will drive away the too inquisitive adult male dolphin, the shark, even perhaps a lone killer whale. In captivity the adult bottle-nosed and common dolphin males, however, normally dominate the other inhabitants of the same species. The mature male is larger and heavier that the female, and having lost the playfulness of youth, will rest much alone, passive, drowsy or watchful, as far from the others as the confines of the tank allow. Here he will receive one or more females during the period when they are ready to mate. He now becomes lively, examining the genital region of the visitor with his snout, but rather than pursue her among the others in the tank if she wanders away, he will call her back.

All dolphins, and indeed all cetaceans observed, have a graceful courtship, arching their bodies as they circle each other, twisting to present the genital region to the amorous partner, stroking with flippers, occasionally mouthing or nipping with teeth, leaping out of the water and diving and splashing. Mating is belly to belly, the male usually on his back; and intromission takes place frequently during the female's receptive period.

Towards other, younger, males, if they approach the oestrous female, the boss bull will clap his jaws noisily, and if the intruder persists, will inflict hard bites. The powerful tail is also a weapon to punish a rival, slapping hard at and perhaps pinning and bruising the other male against the wall of the tank. Towards inquisitive sucklings and weanlings the dominant male is as tolerant as he is towards anoestrous females: he ignores them or at most gives a warning push or nip. In a tank he is an aloof figure; at sea he is likely to be the leader of a pod.

Mature females are next in order of dominance. In movements at sea they seem to keep close to the master bull. In a tank they are livelier and more inquisitive than the old male, although there is considerable individual variation—some cows are shy with their first experience of mating and calving, but gain confidence and rise in the female peck-order as they reach physical maturity and second and subsequent calvings.

Indicative of what takes place in the wild, any sudden invasion of the tank by a newcomer, or a large unfamilar object thrown among them, or a sudden loud continuous noise, will cause dolphins to crowd together, the order of dominance temporarily abandoned; the school will swim in close body contact until the interruption to normal life has ceased or been accepted.

Lowest in the hierarchy of the dolphin pod are the young males and females. They are often pushed aside by the adults. However, they play a great deal together and are easy to train at this age, learning tricks with balls and hoops within a few hours of their first lesson from a keeper. They are also quick to imitate; and in play will invent games of their own, such as balancing a light object on the tip of snout, flipper or tail held above water. Possession of a cork or feather will be competed for. A favourite game invented by a dolphin in one tank was to carry in its mouth a feather or other light object, such as a handkerchief, to the jet of renewal water spurting into the tank, and release it, then chase it, and repeat the experiment. When another youngster joined the game, the pair would alternate roles, one releasing, the other collecting. Sometimes a third dolphin would naughtily intercept and dart away with the prize; a chase would ensue, pursuer and pursued obviously enjoying every moment.

Fichtelius & Scholander in *Smarter than Man?* describe how two dolphins in a tank amused themselves with an eel, each in turn catching the eel, carrying it around for a while with the other in pursuit, then allowing its companion to have possession:

> One day the eel managed to hide in a pipe at the bottom of the tank. The dolphin found a small poisonous fish that also lived in the tank. He carefully caught the fish in his teeth so that it couldn't hurt him, and poked the fish into the pipe where the eel was hiding. The eel zoomed out and the dolphins continued their game of eel catch.

One of the earliest pleasures of young dolphins at Marine Studios was to throw a rubber inner tube on to the trainer's platform. The other young dolphins immediately began throwing a tube at spectators, who pitched it back, and the game continued as long as a human playmate was willing to respond. The young dolphins never tired, and expressed their disapproval with squeaks of protest if the human partner delayed or ceased to return the tube.

These joyous games with portable objects resemble those of puppies and serve the same purpose of exercising and strengthening the body and developing the brain for the serious business of survival, hunting for food, and avoiding predators. It should be noted that these games with human and dolphin partners are initiated for no reward other than pleasure in the activity. Later, as it grows up and becomes a more solemn, or at least less playful adult, and is trained to perform quite difficult feats requiring skill and co-ordination of body and brain, such as high jumps through hoops (perhaps on fire), the trainer finds it necessary to reward the performer with food. Frank Robson, however, has declared that food rewards are not necessary—his dolphins performed out of love of the man.

Bottle-nosed dolphins at the aquarium at Miami, Florida, showed remarkable skill and calculation in their gymnastic tricks, being rewarded by the trainer with a piece of fish quietly slipped into the mouth—opened expectantly—after each successful performance. As I watched I felt that they were eager to perform because they were hungry, but also bored with previous inaction. One had to admire their restraint under these circumstances, patiently watching for the signal from the trainer. I was even more full of admiration when the trainer called one bottle-nosed to play skittles. The ball was thrown into the big pool where this dolphin picked it up in its mouth and swam to the edge of the tank which rose about 40cm (16in) clear of the water. Rearing half its body out of the water, the dolphin paused to stare at the nine skittles placed in the correct position for the game, at the end of a narrow carpet approximately 9m (30ft) from the dolphin. Only a few seconds passed before the dolphin, with a jerk of its head, flung the ball with such accuracy that all nine skittles were bowled over first time. Even an expert human player could not have done better, and he might have done much worse.

This was a surprising and admirable feat of skill and thoughtful concentration for a sea-mammal accomplishing a difficult task on land set by its trainer. Once again I wondered how intelligent the cetacean mind can be. When I spoke to the trainer afterwards he merely laughed and said she sure was a cute lady. I must leave the reader to judge, merely giving one last example of this dolphin working out a problem on its own.

When a feather, tossed repeatedly between a young female dolphin and a ringside spectator, remained stuck on the wall of the tank, the dolphin was unable to pluck it from this position because her protruding lower jaw prevented her from gripping it by a frontal approach each time she reared up and tried. After a thoughtful stare at the feather she solved the problem by brushing it down into the water with a sideways swipe of her snout. One lesson was enough; she knew what to do when it stuck next time.

There is hardly space here to describe more of the large number of intelligence tests devised by man in the training of captive dolphins of several species, but chiefly the bottle-nosed. These include the assistance they can be trained to give to divers working deep below the surface, by carrying tools and other equipment, pressing triggers, even planting adhesive mines and bombs against the hulls of vessels and being skilfully trained to explode underwater mines, thus destroying themselves in the process. Perhaps, fortunately, the fear that a dolphin might place explosives against the hull of the wrong ship, plus the public outcry in the United States against their use for this purpose, has deterred the regular employment of dolphins for naval warfare. Research continues, some successfully directed to teaching dolphins to rescue humans in distress at sea. It has even been suggested that these humane and clever friends of man could be trained to patrol along the surf line of popular bathing beaches and drive away man-eating sharks. Dolphins have been seen attacking large sharks swimming too close to a family pod by repeatedly butting them in the gills and so wounding them in this way that the great fish has subsequently died.

Common Dolphin *Delphinus delphis*

This is the species depicted in the stylised but beautiful paintings, and on the early coins of the Mediterranean nations, along whose coasts this dolphin has ever been numerous. The Dolphins Fresco in the Palace of Knossos, Crete, dates back to about 1600 BC; as restored by Gillieron it is a perpetual delight to gaze upon, so wonderfully alive are these *Delphinus* in their marine environment, with different species of local fish and the suggestion of the waving sea throughout.

About a dozen subspecies have been named, covering distribution throughout the world in waters not colder than 15°C. Its general life history is described under the last species. At sea it can be identified by its smaller size, rarely exceeding 2.5m (8ft) and up to 130kg (290lb). It has a darker back, and a sharp black line which begins at the upper jaw and curves above the eye towards the dorsal-fin, then sweeps to the genital region, behind which the rest of the body is dark. The belly elsewhere is white, and in action the fin has an identifying white flash, clearly visible when this dolphin leaps clear, as it does very freely. Body colour is at its brightest during the summer breeding season when mating takes place and the calves are born. Gestation is just under one year.

In the attractive Whaler's Cove pool at Sea Life Park, Hawaiian Islands, this dolphin has been studied in the company of the next (*Stenella*) species. The two mate freely in captivity, but so far without resulting offspring. The largest male, a *Stenella attenuata*, dominated one female *Stenella* and two female and three male *Delphinus;* these six others each had their respective positions in the hierarchy. The order of dominance was maintained even in the sleep formation, when the mixed pod circled the pool slowly, resting near the surface and periodically rising to breathe. Indicative of what happens in the wild, these dolphins were never quite fast asleep, but as they very gradually circled the tank, both eyes of each individual did not shut for more than one blink; as far as could be ascertained one eye would be opened for up to ten minutes; when it closed the other eye was opened. Evidently in this way the dolphin obtains sufficient rest, by a half-sleep, with one eye always on watch for danger.

Studying this particular mixed group Dr Gregory Bateson, author of *Steps to an Ecology of Mind*, after many years of research into cetacean intelligence, concluded (like John Lilly after trying in vain to work out a delphinese language) that too much emphasis has been placed on trying to tell the captive dolphin what to do, rather than let it show us how it determines its problems naturally. A few months ago at Whaler's Cove, watching these beautiful animals, I too felt that we are too anthropomorphic. We forget that cetaceans live not by their hands as we do, but by their 'seeing ear', by sounds and ultrasonic communication which divine in depth through water, through bodies, in a very different and, to us, alien medium. As from time to time I watched the eye of dolphin,

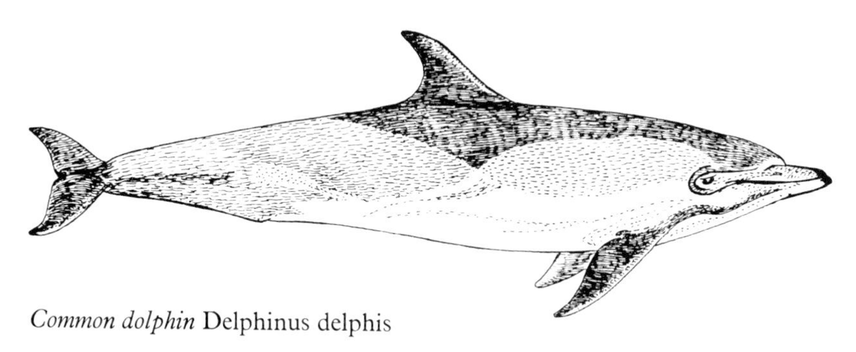

Common dolphin Delphinus delphis

pilot and false killer whale flash above water to give me a passing stare (the performance for the day had long been finished and the whales were behaving as quietly and naturally as they might in this large pool from which they could hear the noise of the Pacific surf only a few yards away) I felt I could almost put into words their thoughts. They were somewhat melancholy—at least I came away with a feeling of sadness when I left them, so close to the freedom of the sea.

'You have seen what we are capable of,' I seemed to hear one old friend, a false killer, signal me. 'Is it not enough that you have removed us from our joyous freedom in the ocean (for we are world travellers, we *Pseudorcas*) and placed us in this miserable prison where we are tormented by the roar of the surf just below our walls? That we have consented to be docile and show you how intelligent we are in interpreting the peculiar demands and successfully carrying out the devious tests you make of our skills in water and air? What more do you want before you set us free? We are very tired and desperately bored with you gaping humans, and with the tiny swimming space you think is such a splendid home for us to live, and die. As one intelligent form of life to another I warn you, you will get no further in understanding the cetacean mind until you swim side by side with us in the free society of the illimitable ocean.'

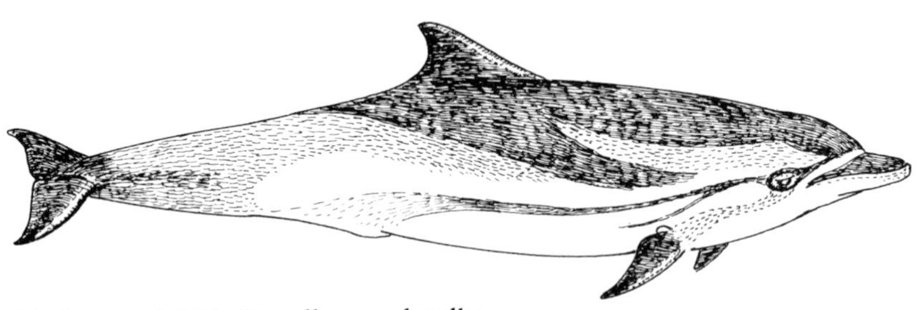

Euphrosyne dolphin Stenella coeruleoalba

Stenella **Dolphins**

Again a large group, of almost identical appearance and size with the *Delphinus* genus, but separated because *Stenellas* have a deep groove in the palate lacking in other dolphins. They are less vividly coloured, mostly brown above and pale below, with the rear and whole of the flukes the same colour as the back. Taxonomists still wrangle over separation into species and subspecies. Unfortunately for the observer, puzzling over identification at sea, distribution is not much help, since all groups are represented in the four major oceans. With luck these groups can be distinguished as follows: *S. longirostris* has a long slender beak, more uniform brown. Probably more coastal than others. *S. microps*, somewhat larger, up to 2.1m (7ft) and 90kg (200lb), is considered to be strictly North Pacific. These are the spinners, so-named because they spin around during their joyous aerial leaps clear of the surface. Spotted dolphins *S. attenuata* are darker, but again with some variation. Most, sometimes all, of the body is heavily spotted with white, but this is not easily seen unless these dolphins swim close. The beak is long. Oceanic and coastal. The euphrosyne or striped dolphin *S. coeruleoalba* is dark blue to grey white, with white on the belly. A narrow dark lateral band extends from the eye to the anus. Mainly oceanic.

PORPOISES

The smallest cetaceans are included in this sub-family *Delphininae*, dolphins with a blunt head and no obvious beak. As a rule they soon die when placed in captivity so little is known about their life history except what can be gathered from post-mortem examination. They are mostly coastal, freely entering estuaries and harbours, often in very large schools. They are difficult to shoot, being swift and lively, and jump clear somewhat less freely than the small dolphins already described. However, fishermen take them in nets and they provide a regular seasonal fishery in many parts of the world, chiefly on the more remote coasts and islands.

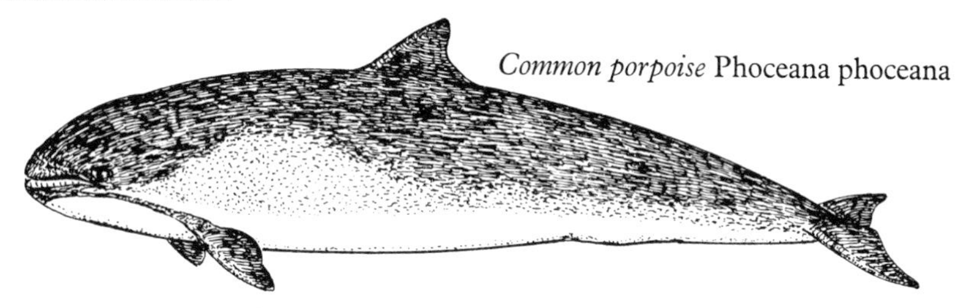

Common porpoise Phoceana phoceana

Common or Harbour Porpoise *Phocaena phocaena*

This is abundant in the temperate and cool coastal seas of the northern hemisphere. I have seen large schools along the coasts of the British Isles, Denmark, Iceland, Madeira, Florida and Maine. Length averages 1.8m (6ft) and weight 75kg (165lb). Typically it is dark above, with a dark stripe from the eye to the flipper, and clear or dull white below. Its meat was eaten in medieval times in Britain and Catholic countries on Fridays, pronounced by the priesthood as fish for that meatless day. At that time the Crown claimed all porpoises and dolphins as royal property. This porpoise leaps well but in normal breathing the blunt snout scarcely appears; the brief blow is followed by the split-second flash of the black dorsal-fin. The race *relicta* is smaller, inhabiting the Black Sea, and has been overhunted to such an extent that it is now protected in USSR waters and is slowly recovering, though rare and confined chiefly to the Sea of Azov. The porpoise of the warmer water of the coasts of California south to Panama has been given specific rank as *P. sinus;* it is a little smaller and lacks the eye-to-flipper stripe, is greyer, and paler below. Although it has rarely been examined by scientists, fishermen know it well from capture in their nets.

Spectacled Porpoise *Phocaena dioptrica*
This striking porpoise, sharply black above and white below and around the snout and tail, is found from Patagonia and the Falkland Islands to South Georgia. It is slightly smaller at 1.5m (5ft) than the southern right whale dolphin which has a similar distribution and piebald pattern, but the latter has a pronounced beak and no back-fin.

Black Porpoise *Phocaena spinipinnis*
Coastal waters of southern South America, seldom reported, all black, rather like a stumpy false killer whale with a small flattened dorsal-fin, and only 1.5m (5ft) long.

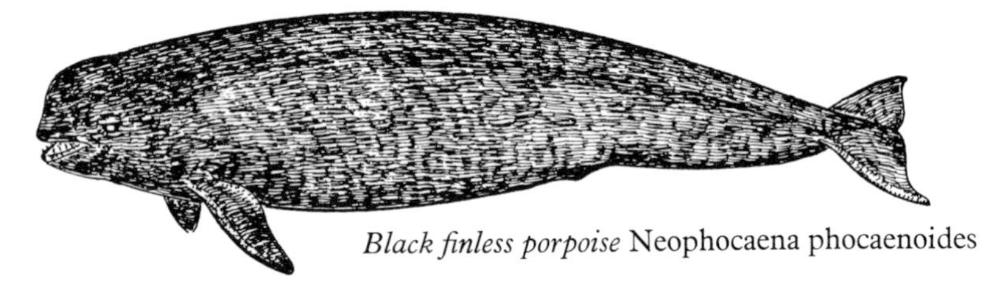

Black finless porpoise Neophocaena phocaenoides

Black Finless Porpoise *Neophocaena phocaenoides*
In place of the dorsal-fin this all-black porpoise has a dent followed by a sharp ridge, hardly perceptible at sea. It is small, up to 1.5m (5ft), lively and slim. Found on the east Asia coast, it has been seen 610km (1,000 miles) up the Yangste Kiang River in China; in cold water about 5°C off Japan; as well as near tropical coasts of India and in Bombay Harbour where the sea temperature is around 28°C.

Dall's harbour porpoise Phocaena dallii

Dall's Porpoise *Phocaenoides dallii*
Up to 2.1m (7ft) long and 136kg (300lb). Numerous in the North Pacific and recognisable by its white flanks. Often plays in the bow-wave of ships and accompanies whales. The tip of the sharp dorsal-fin is sometimes white. Migrates between California and Japan and sub-Arctic coasts. Like other true porpoises, and most dolphins, the calf is born in the summer and is well-grown on the return to winter quarters.

True's Porpoise *Phocaenoides truei*
The white extends above the flipper almost to the eye, is probably a

subspecies of *P. dallii*, and is found with it in Japanese waters. The horny projections between the gums may assist in holding slippery squid food.

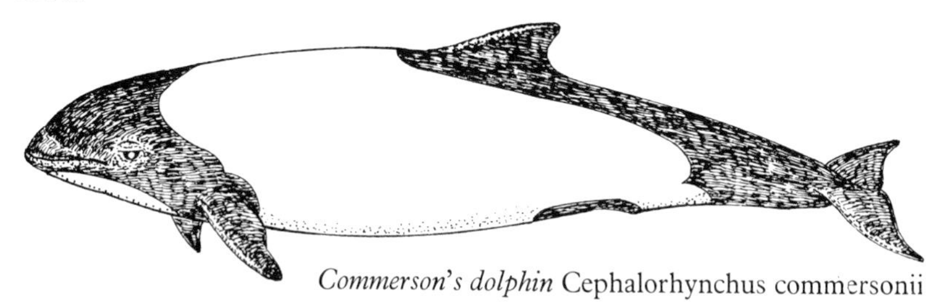

Commerson's dolphin Cephalorhynchus commersonii

Sub-Antarctic Porpoises or Dolphins *Cephalorhynchus*
Between 1.2m (4ft) and 1.5m (5ft) long, blunt-nosed and with blunt dorsal-fin, these small southern species haunt coastal waters far from the main shipping lanes, except *C. heavisidei* which is frequently encountered off the Cape of Good Hope. They are pied, with varying amounts of white and black. *C. commersoni* has a white body, black head, fins and rear; and ranges from Patagonia to the sub-Antarctic south of the Atlantic and Indian Oceans. *C. hectori* has much the same bicolor pattern but with a white or pale forehead; it is often numerous in shallow water in New Zealand where it can be too tame, getting into fishermen's nets, even swimming among human bathers—it has been filmed playing with children in Golden Bay, South Island. *C. eutropia*, with hardly any white except along the belly, is apparently confined to Chilean coasts.

Ploughshare Dolphins *Lagenorhynchus*
These are almost twice as large as the last group and comprise six species, pied, with white bellies and carrying a tall dorsal-fin. They are fast swimmers and lively leapers. The white-beaked *L. albirostris* is the largest, up to 3m (10ft) long and 270kg (600lb). With the rather smaller white-sided *L. acutus* it ranges the temperate waters of the North Atlantic Gulf Stream, the coasts of Maine to Iceland and the North Cape of Norway. As they fail from age or sickness these dolphins are frequently washed ashore on these coasts. The tall fin of the Pacific white-sided representative *obliquidens*, up to 2.5m (8ft), is half white (rear portion),

White-beaked dolphin Lagenorhynchus albirostris

and it carries more white on its flanks, although this is variable in the way it is patterned.

White-sided dolphin Lagenorhynchus acutus

Lagenorhynchus crusiger

Hourglass Dolphin *Lagenorhynchus cruciger*
Smallest of the ploughshare dolphins, up to 1.8m (6ft), having two striking wavy white panels, faintly resembling the two globes of an hourglass, between the nose and the tail. The back, dorsal-fin and flippers are black, the throat and belly white. It lives close to the Antarctic ice and consequently is rarely seen. It is replaced in latitudes 50° to 30°S by the dusky dolphin.

Dusky dolphin Lagenorhynchus obscurus

Dusky Dolphin *Lagenorhynchus obscurus*
This is a good deal larger, up to 2.5m (8ft) and 136kg (300lb). The dusky has a similar bicolor pattern, but the white is always dusky, and sometimes so dark that it may appear black at a distance. The New Zealand coast is a favourite haunt, and here it has been successfully kept in

captivity and studied by Frank Robson. Robson considers that age groups go about in separate schools.

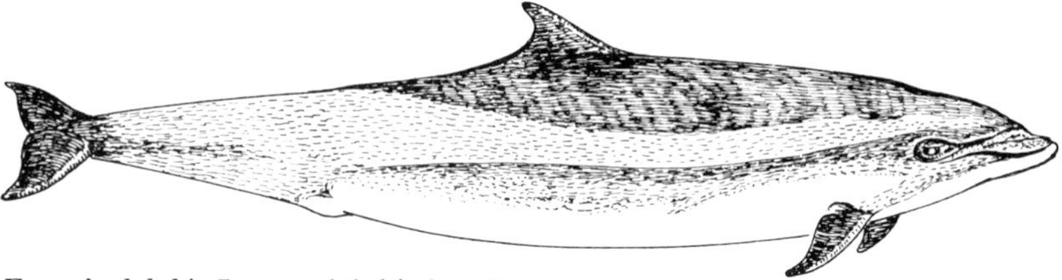

Fraser's dolphin Lagenodelphis hosei

Fraser's Dolphin *Lagenodelphis hosei*

Originally described from a skull at the British Museum (Natural History) in 1956 by Dr F. C. Fraser which had been collected from a beach in Sarawak in 1895. Its appearance remained unknown until specimens were collected by Best, Gambell and Ross in an area south of Durban in 1971. A dark greyish-blue dorsally with a characteristic dark stripe from the head to the anus, which has obviously led to confusion in the field with the euphrosyne dolphin. It is interesting to note that a new 2.4m (8ft) mammal can be described for the first time in the 1970's. Captain Bruyns (1971) has observed closely and sketched two unidentified beaked dolphins in the South China Sea and the Malacca Strait, which he has tentatively included in the same group in his book. The number and variety of beaked dolphins (*Delphinus* and *Stenella*) is bewildering, even to his expert eye.

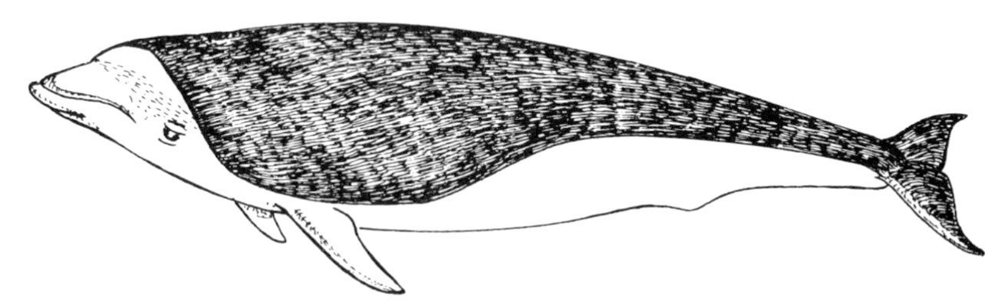

Right whale dolphin Lissodelphis peronii

Right Whale Dolphins *Lissodelphis*

These are much easier to identify; they have no dorsal-fin and are very swift and graceful, darting through water and air like arrows. The northern form *L. borealis*, up to 2.4m (7.9ft) and 90kg (200lb), black with a white belly stripe, inhabits the North Pacific from California and Japan, migrating north to summer in the Bering Sea. It has been seen moving in a school close to those of white-sided dolphin and pilot whale. The southern *L. peronii* is smaller, up to 1.8m (6ft), and rarely seen as it ranges all sub-Antarctic seas: the white extends to the whole of the face beyond the eye and the tail and fins are white.

Rough-toothed Dolphin *Steno bredanensis*
A fast swimmer and leaper, up to 2.4m (7.9ft) long, not unlike the last but with a prominent pointed back-fin. So-called because the conical teeth are serrated. The upper parts are black, spotted with white stars, the white under parts suffused with pink and sprinkled with dark spots. A sub-tropical and tropical dolphin found in all oceans, with strandings as far north as Holland. Mass strandings have occurred in Senegal and Florida. Bruyns considers that a smaller *Steno* species is abundant in the north-western Indian Ocean, but so far this is unconfirmed.

Bufeo or Amazon Dolphin *Sotalia fluviatilis*
Also known locally as Tookashee, this is a true delphinid, separate from the four river platanistids. *S. fluviatilis* is, however, seen in company with the Amazon river dolphin known as the bouto (page 184) and ascends this great river for 2,500km (1,500 miles). One-third smaller than the more numerous bouto, it is dull-coloured, grey above and whitish below, and rather shy of man. The bufeo seems to prefer shallower water and will roam miles through inundated forests during floods, whereas the bouto keeps more to mid-river and roomy lakes. *S. guianensis*, slightly larger, is more coastal with a range covering the whole shore and estuaries of Brazil north to and including Guyana. Both have short dorsal-fins and medium beaks.

Sousa teuszi
Little is known about this dark-coloured long-beaked dolphin which haunts the rivers and estuaries of tropical West Africa. The back-fin rises a few inches above a low ridge almost one-third the length of the 2.2m (7.2ft) body.

Plumbeous Dolphin *Sousa plumbea*
Up to 2.8m (9.2ft) long, and like the last has the same curious long dorsal ridge surmounted by a small fin. Grey-coloured with lighter spots, it ranges the tropical Indian Ocean coasts from Natal to the Red Sea and the Persian Gulf. Frequently observed in harbours, once seen to hide behind the hulls of ships from whence it dashed forth to surprise its fish prey. Further east the smaller Speckled *S. lentiginosa*, curiously blotched with white and pink panels, occupies muddy rivers and silt-laden estuaries in the Bay of Bengal. It has a smaller dorsal ridge and fin.

Borneo White Dolphin *Sousa borneensis*
Pale, white, sometimes flush pink, this conspicuous small dolphin—up to 1.8m (5.9ft)—haunts the north shore and rivers of Borneo. Further north the Chinese White *S. chinensis* in the Gulf of Siam and nearby Chinese estuaries is so alike in size and colour, with some ivory and some rose-tinted individuals, that it may be a subspecies. Both have the same small ridge-fins.

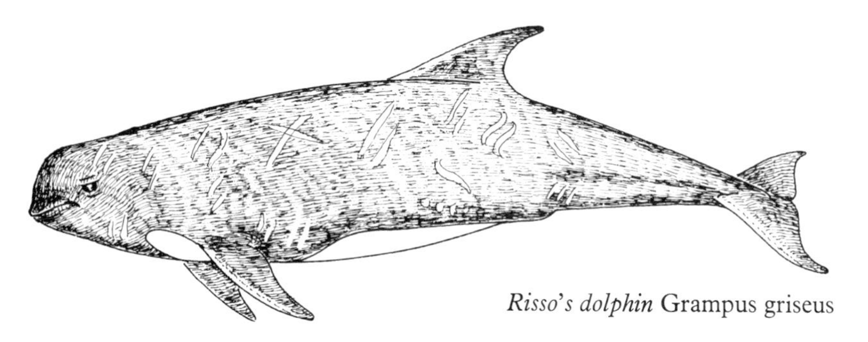

Risso's dolphin Grampus griseus

Grampus or Risso's Dolphin *Grampus griseus*
A large blunt-nosed dolphin up to 4m (13ft) long and 680kg (1,500lb) in weight. Generally a dull colour, it grows paler at maturity and can be almost white with age. The fin is tall, 40cm (16in), and curved backwards. Usually there are many white lines and scribblings on the upper parts, suggesting aggressive body contacts. It is widely distributed except in polar seas but is local. One such famous dolphin, Pelorus Jack (page 29) was protected by an Order-in-Council in New Zealand, where this species has not yet officially been satisfactorily identified! It is rare along the west coast of America but common along the coasts of China as far south as the Coral Sea, and in the Indian and Atlantic Oceans. I have watched a small pod with calves feeding and leaping in the Minch, Inner Hebrides, where fishermen know it as the Lowper to distinguish it from the blackfish or pilot whale, both common there in summer. In September both appear off the Welsh coast where I have seen schools passing south, probably to winter quarters in or south of the Mediterranean. In captivity both species show typical dolphin ability to learn tricks rapidly.

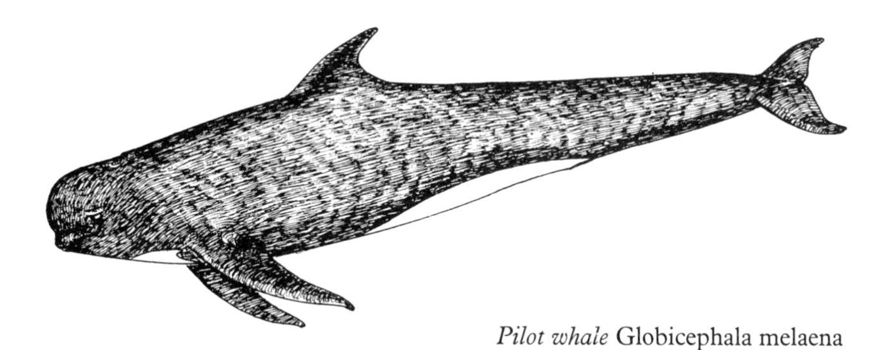

Pilot whale Globicephala melaena

Pilot Whale or Blackfish *Globicephala melaena*
The huge melon grows more bulbous with age, especially in the male, which at 8.5m (28ft) is one-quarter as large again as the female and weighs up to 3,800kg (8,400lb). The colour is very black, except for a white breast-plate between the flippers, tapering to a narrow white line

beyond the anus in the North Atlantic race. This white belly streak is grey in the southern and Pacific races. The flippers are long, narrow and pointed, the dorsal-fin prominent and well forward. This whale ranges all seas with temperature above 13°C, usually in tight regimented formations of up to 200, surfacing with a loud visible blast 1.5m (5ft) high. The calf is born late in the spring and probably for more than a year remains close to its mother in the family group, which seems to be dominated by the largest male, as in the cachalot. Their migrations seem to be coastal, or at least in waters of the Continental shelf, but have not been at all well worked out. They are quite indifferent to passing ships, hardly moving out of the way.

Now and then quite large pods strand upon a shallow beach naturally, as when forty came ashore near the gannet colony at Cape Kidnappers, New Zealand, in January 1967. This occasion was notable for the successful refloating of all but four of the heaviest; although they appeared to be dying, when each one was held upright in a few feet of water, and slapped and encouraged to regain its sense of balance, it revived and swam safely back to sea, and the pod departed, leaving only four dead behind. From the habit of cruising through narrow sounds in large schools the pilot whale has been hunted from time immemorial. Small boats form a half-circle to head the whales towards a shelving beach. As soon as the school is moving towards the cul-de-sac, the hunters beat the water with oars and sticks, making such a noise that the whales lose their sense of direction (sonar guidance) and rush from the pursuers. The first to strand and call in distress is the signal for the rest of the school to pile up on the shore beside it. Faroese islanders rely on killing several hundred annually; the carcases are divided between participants and occupants of the nearest houses, and the meat processed for winter use. In Newfoundland 3/4,000 are taken by small whaling boats annually. A school of thirty swam up the Thames in November 1965 as far as Tilbury Docks.

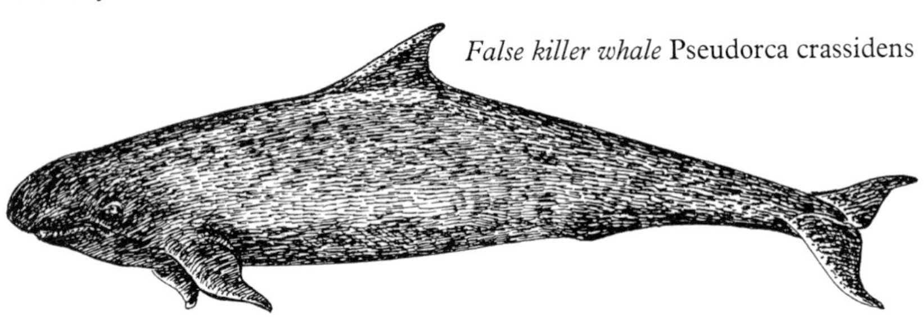
False killer whale Pseudorca crassidens

False Killer Whale *Pseudorca crassidens*
An all-black world roamer of temperate to tropical seas, so-called because it is shaped like an orca, with an array of strong conical teeth, 20 in each jaw, but it is altogether smaller and slimmer. Males reach 6.1m (20ft), females 4.6m (15ft); weight up to two tonnes. Strandings are large at times: 253 died in one recorded in New Zealand (page 10); these

had light white scribblings, not deep scars, rather as if some snail had crawled over the black skin, leaving a permanent trail. Their wandering migration is little understood; they may appear in a large school in the deep ocean or in very shallow water. Their teeth ivory was an item of trade wherever they were killed, or stranded, and is still valuable for the carving of small objects. This whale is very common in the north Indian Ocean, where large family pods have been seen with small and older calves playing close to the large tankers in the Persian Gulf. It was a pleasure to watch their friendly behaviour in the Whaler's Cove sea-quarium in Oahu, Hawaiian Islands. But we know little about their breeding biology.

Little Killer or Broad-beaked Dolphin *Peponocephalus electra*
A slim small edition of the last, dark above, a little paler below, and with whitish lips. Up to 2.8m (9.2ft) long. Although it is recorded from all tropical seas, it is rarely seen, and its life history unknown.

Pygmy Killer Whale *Feresa attehuata*
Resembles the last in colour and shape, but smaller at 2.4m (7.9ft) Very rare, a few records only, in tropical Pacific and Atlantic. Fourteen captured in Japanese waters and placed in the Honshu Aquarium died within 22 days. In 1965 one of a school of 50 near Hawaii, taken to Sea Life Park, showed aggression by flipper and fluke slapping at the surface, growling and jaw-snapping. A pilot whale in the same tank was repeatedly butted in the throat region and was found dead next day.

Irrawaddy Dolphin *Orcaella brecirostris*
This curious-looking dolphin is dark grey and about 2m (6.6ft) long. The head bulges like that of the pilot whale, but the dorsal-fin is stumpy and the flippers short and broad. It inhabits warm coastal waters between Calcutta, Indonesia and Borneo, and skulls have been found in northern Australia. It has been seen 1,450km (900 miles) up the Irrawaddy River in Burma; its round head and small dorsal-fin are easy to identify, yet little is known about its breeding habits.

Platanistids **Freshwater Dolphins**
Last to be considered, but often first on the list of the toothed cetaceans because they exhibit some primitive features. They are slow-moving and hard to watch in their muddy riverine habitat, and one is quite blind. In captivity they appear as intelligent as any ocean-going dolphin. They are remarkable for their more supple neck which enables them to turn the head up, down and sideways, because the neck vertebrae are not fused together rigidly as in all other dolphins except the narwhal and beluga.

They are in a sense the most terrestrial of dolphins, keeping very close to the bottom of rivers and lakes, and even threading their way through flooded jungle. They retain more hair about the face than other cetaceans, but it is highly specialised in the form of stiff bristles on the very long thin snout, evidently used, like that of the cat and seal, as a tactile guide in the darkness of their environment.

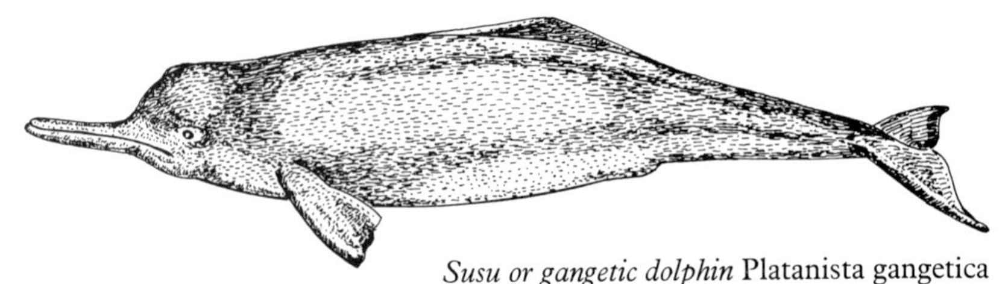

Susu or gangetic dolphin Platanista gangetica

Susu or Gangetic Dophin *Platanista gangetica*
This dull-grey dolphin has no lens in the tiny eye and is blind. It navigates expertly, however, and locates its food by sonar and its sense of touch through snout and flippers. Living in the muddy waters of the Ganges, Bramaputra and Indus rivers of India, it has little use for sight. The susu is rather longer than a man, up to 2.5m (8.2ft) and weighs about 73kg (160lb). Where it frequents the sacred waters of zones set aside for Hindu religious observances it is tame and glides about close to the legs of human bathers, apparently confident that it will not be molested. It has been found 1,600km (1,000 miles) up river. Pregnancy lasts nine months; the calf is born between April and July. The fins are rectangular paddles, said to assist in reversing movements. They are also used to clasp each other, belly to belly, during mating; there is a lively courtship as these blind dolphins leap and caress each other at the surface; finally the amorous couple rise vertically in each other's 'arms', and as their genitals meet the male thrusts into the female who falls gently backwards; still locked in coitus the pair vanish beneath the surface.

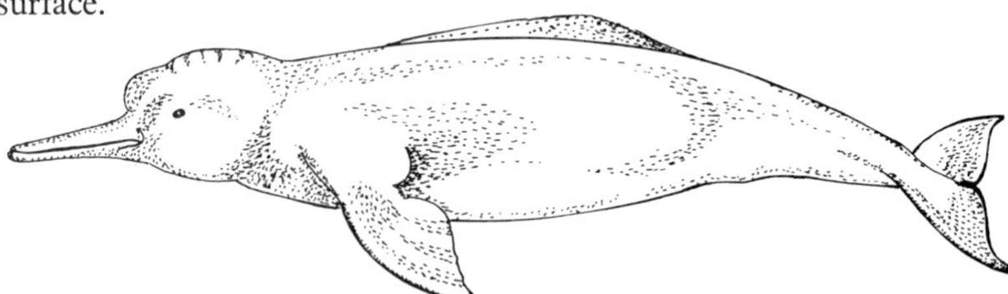

Bouto or Amazonian dolphin Inia geoffrensis

Bouto or Amazon River Dolphin *Inia geoffrensis*
Not to be confused with the bufeo dolphin *Sotalia* with which the bouto associates in the same river. The bouto is larger, up to 2.8m (9.2ft) and 110kg (240lb), and more numerous, with a longer beak and only a ridge in place of the bufeo's dorsal-fin. A dark animal, paler and sometimes rosy below, it has small but good seeing eyes. The local name comes from the sound of its blow. Limited to the Amazon and Orinoco and tributary rivers, from the estuary to many hundreds of kilometres in-

land. Amerindians will not kill it, believing this would bring misfortune, and blindness if the oil is used as lamp fuel; also they use this dolphin to help catch fish (page 42).

During tropical rainstorms this dolphin may leave the main river like the bufeo does and pursue its food through the flooded maze of the jungle. Its accurate sense of orientation enables it to return afterwards through the muddy shallows where a man without instruments of navigation would be quickly lost trying to find the way under the canopy of tall trees.

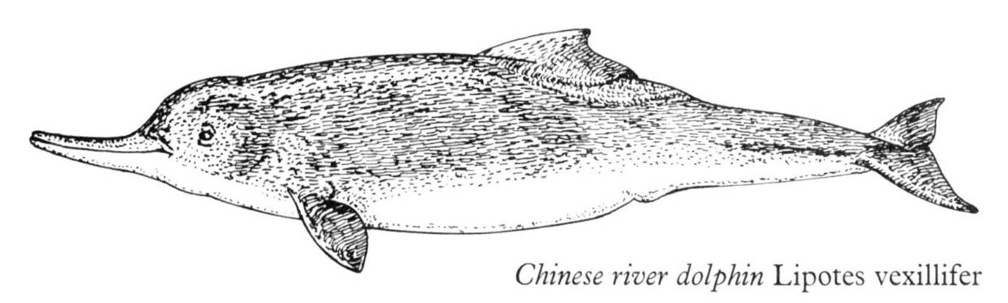

Chinese river dolphin Lipotes vexillifer

White Flag or Chinese River Dolphin *Lipotes vexillifer*

The pei-chi, about 2.3m (7.5ft) and 80kg (180lb), has been little studied. It is isolated, some 900km (600 miles) from the sea in the Yangtse River and the Tung Ting Hu Lake. This grey-white dolphin with the up-tilted beak is said to breed in the wet season in the upper waters of tributary rivers and in the dry season to herd together in shrinking pools in the main river and lake when the white flag of its dorsal-fin gives its position away. However, it is not normally hunted; a Chinese fairy tale associates it with the drowning of a princess.

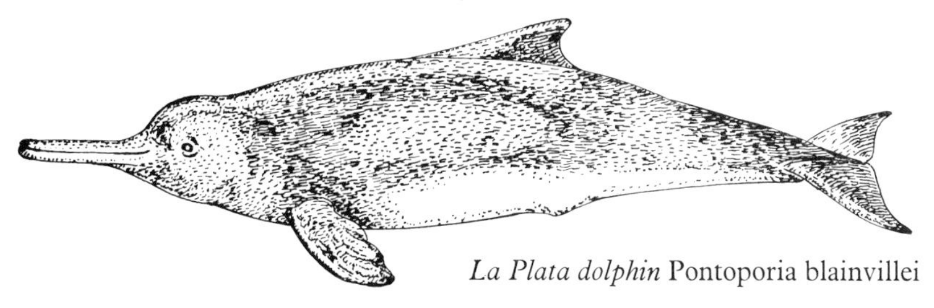

La Plata dolphin Pontoporia blainvillei

La Plata Dolphin *Pontoporia blainvillei*

Pale brown above, whitish below, this small dolphin with males to 1.5m (5ft) and females to 1.7m (5.6ft) is found in the lower reaches of La Plata River, Argentina, and probably spends part of its annual cycle in salt water. Fishermen take them in nets, but almost nothing is known of the breeding cycle.

Southern right whale breaching (Jen & Des Bartlett/Bruce Coleman)

Saving the whales

As you read this, the slaughter of whales and dolphins continues. Every twelve minutes a whale is killed—the living tissue blown into agony by explosive harpoons. LET US ACT.

Joan McIntyre (Project Jonah 1974)

In *Mind in the Waters* Joan McIntyre reminds us that whaling is not something out of a romantic past, but a highly mechanised industrial operation of the modern world. While scientists discuss whether or not there should be a moratorium on whaling, factory ships and purse-seine netters ply their grisly trade for personal profit through the oceans of the world.

And what of the future? For the large baleen whales it is bleak, and for the small whales and dolphins not at all bright.

The state of world population of commonly hunted species is taken from the latest report of the International Whaling Commission, with some amendments and additions to bring it up to date as this book goes to press.

Species and Original Numbers 200 years ago (estimated)		Present numbers
Blue	200,000	1,200
Fin	500,000	150,000
Sei	250,000	170,000
Right	100,000	5,000
Minke	250,000	150,000
Bryde's	50,000	40,000
Gray	50,000	11,000
Bowhead	75,000	2,800
Humpback	250,000	6,000
Sperm	2,000,000	1,250,000
Narwhal	60,000	30,000
Beluga	60,000	30,000

World numbers 200 years ago can only be guessed at; the original populations of great whales living in equilibrium with their environment, as yet not seriously predated by man, can never be accurately known. The present-day figures given above have been arrived at from a combination of published statistics of the whaling industry (numbers, ages and sex of those killed), and from sightings at sea; they are believed to be reasonably accurate estimates for the purpose of assessing the Commission's proposals for conservation or exploitation of surviving stocks.

The only whales given total protection world wide as I write this are the blue, the humpback and the right (except for a limited number of Greenland right whales taken by Eskimos); but even these are at risk from certain pirate ships flying the flag of a nation which is not a member and does not accept the ruling of the IWC. Among conservationists there is real fear that the surviving pods and individuals of the rarer whales which live isolated from each other by their habit of keeping to local feeding and breeding grounds will fail to rear enough immatures to replace the ageing breeders and establish stronger units. In nature there is a threshold of numbers below which the species is no longer viable.

The only hope is to achieve quickly universal conservation by cooperation between all maritime nations. Pressure must be brought to bear on those nations which still allow indiscriminate hunting. There is also an urgent need to bring all cetacean species into the range of international regulations to prevent over-exploitation. The smaller whales and dolphins have always been hunted by local fishermen, notably in the Pacific and the Mediterranean (including the Black Sea), but always in small numbers, an almost natural predation, without seriously depleting the species. But with the introduction and now almost universal use of the purse-seine net, many thousands of the smaller cetaceans are being caught and killed daily, either deliberately or by accidentally becoming entangled in these lethal nets.

In this method the fish shoals are first located by asdic sonar—the dolphin's way—and the dolphins may already be feeding on the shoal. The net, which may be several hundred metres long, is rapidly brought in a circle to enclose the shoal. The foot-rope is drawn tight, and all that cannot escape through the mesh (designed to hold the edible fish) are trapped. Some dolphins leap over the buoyed top of the net, but many in panic get entangled and drown—and these include small whales and porpoises. The fishermen are ill-pleased when the more powerful animals break through the nets, perhaps carrying a portion of the tough nylon around their bodies. I have seen sea-lions at the Pribilof breeding grounds in the Bering Sea draped with sizeable fragments of these nets. and if they are unable to free themselves they eventually die from this impediment to normal feeding.

Although the purse-seine net is an enormously profitable investment for the fishing industry, it is also costly, and it is not surprising that fishermen will shoot or otherwise slaughter dolphins, seals and other large animals which are liable to tear holes in the mesh. Normally the bodies of these are discarded when the seine is hauled and the edible fish stowed aboard. But of recent years several fishing companies have found it more profitable to stow everything that comes aboard in tanks in the hold. Often one haul of the purse net will fill the tanks of two seine fishing boats (usually working in pairs); the contents, alive or dead, dolphin and fish, raw and unrefrigerated, are dumped into the hold and rushed to the company's shore station factory or the accompanying floating factory ship. Here the whole cargo, including any small whales, dolphins, seals and sharks, is unloaded by hydraulic machinery and processed into high-priced fish meal and fertiliser sold for livestock and agricultural consumption.

The purse-seine nets are in fact a greater disaster to cetacean survival than present-day whaling, which has fortunately dwindled, although lamentably because whale stocks are too low to be worth the expense of fitting out and maintaining the specialised ships. At present only Japan and Russia conduct pelagic whaling.

Conservation groups all over the world, notably Greenpeace, Friends of the Earth, Project Jonah, supported by the World Wildlife Fund and other wildlife savers, have called world attention to the plight of the whales and dolphins. Money has been found to pay for research into devising release panels and doors to be fitted to the deadly seine-nets, and these modifications seem to be the only way in which, hopefully, the cetacean mortality from this source can be reduced. Meanwhile the slaughter continues; fishermen and fishing companies amass fortunes, while the natural food chain of the sea is harvested and depleted with gathering momentum.

Added to these threats to cetacean existence is the increasing pollution of the sea from many sources: dumping of radio-active waste; heavy metals (mercury, zinc, iron, cadmium); plastic, oil slicks. All these seriously affect the zoo-plankton and thus the whole food chain.

The vanishing life of the oceans is a terrible indictment of man's stupidity and inhumanity towards his fellow creatures, and to their and his posterity.

The International Whaling Commission

This is the only international body officially supported by membership contributions of the major and many of the smaller nations for the regulation of the whaling industry. At first it had little to do with conservation: it was born out of the need to preserve whale stocks for commercial exploitation as a result of gross overcatching in the Antarctic. In the 1930–31 season 41 pelagic factory ships and 6 land stations, with 232 catching vessels, killed 29,410 blue, about 10,000 fin and smaller numbers of sei and humpback whales. This catch yielded 3,600,000 barrels of oil. Not surprisingly the market was flooded, depressing the price to such an extent that the major whaling companies of Norway and Britain agreed to restrict the catch in future to restore profitability.

The British government had long been aware of the threat to whale stocks earlier than this, and had set up the *Discovery* investigations based on the Falklands and South Georgia Islands, where scientists made those studies of the bodies of whales brought to the processing stations which are the basis of our present knowledge of the anatomy of baleen and sperm whales. The Norwegian government at the same time set up a Bureau of International Whaling Statistics, which has continued to collect and publish information on the species, numbers and measurements of whales caught annually.

Oils and fats were in short supply after the war and whaling was renewed during the period of economic recovery when a number of nations took part in a convention for the Regulation of Whaling, signed in Washington in 1946. This was designed not so much to conserve whale stocks believed to have been built up during the near-cessation of whale hunting 1939–45, but to revive the industry on an orderly basis and to provide consumers of whale products with the raw materials in economic demand. However, the Convention established the present International Whaling Commission which has struggled on ever since. Its history makes sad reading; of resolutions passed but not observed; of quotas limiting the killing of the rarer species, but exceeded. The weakness of the IWC lies in its voluntary status; it has no power to enforce its regulations, and even a paid-up member nation can dissent from a majority decision by entering an objection within 90 days of formal notification, so exempting that nation from the obligation of the regulation. But if this loophole had not been provided it is doubtful if the IWC could have continued to exist.

Nevertheless there is today a hopeful sign, in the increasing number of nations joining the IWC; almost all the maritime nations are now members. It is certain that public opinion on world conservation issues has had a considerable effect in achieving this greater world interest in the protection of whales and dolphins. Through television, radio and the mass media generally there is continuing pressure to implement the ten-year moratorium on all whaling first proposed at the United Nations Conference on the Human Environment at Stockholm in 1972.

Such a moratorium for the large whales has become imperative; but just as urgent is the need to save the little whales and dolphins from the slaughter of the fisherman's modern nets.

Today the International Whaling Commission is emerging as a much stronger organisation for the conservation of the cetaceans. It is still only in effect a debating society, but it has compiled a detailed programme for an International Decade of Cetacean Research, to be funded by international organisations, on the results of which research future conservation will be more securely based for action.

In 1976 the IWC established a new secretariat with headquarters in Cambridge, UK. With more nations participating, there is now a better chance of implementing the recommendations of the IWC's annual meeting and, through the presence of representatives of international conservation groups, to implement stronger protection measures.

All is far from well, however. In the words of Dr Ray Gambell, whale research scientist, wise conservationist and the present Secretary of the IWC:

> Some people claim that the achievements of the IWC over the past 28 years are insufficient for the needs of the present time. They argue for a completely new organisation which would be responsible to the world community for this largely international high seas resource. They question the basis and accuracy of the scientific assessment on which the management policy of the IWC is built. The concept of MSY (maximum sustainable yield of numbers of a hunted species which can be killed annually without reducing population size) is held in doubt. The stated objectives of management, of harvesting and utilisation of this natural renewable resource are disputed and alternative objectives which do not require whales to be killed at all are held in their place.

I take this to mean alternatives to the products of whaling. These are already available in the present shortage of whale products, alternatives which will be in even greater demand if the hoped-for moratorium on whaling is ever realised. A complete useful list of such alternatives is given in the Whale Manual 1978, published by Friends of the Earth, London. From this it is plain that we can indeed do without killing any whales, and it looks as if we shall have to in the not too distant future if the present rate of cetacean slaughter continues. As to a brand new organisation to be responsible for the future of the whales, no, we must hurry on with supporting and guiding the present one which has taken over thirty years to reach its new strength. There is no time left to conceive and gestate (and perhaps abort) a new child of the present marriage of whaling and anti-whaling parents.

Dr Ray Gambell concludes (*Marine Policy*, October 1977):

> It is in the context of this ongoing discussion that the present role of the IWC as a forum for debate and negotiation should be judged. The Commission needs to satisfy all concerned that it has the will and capacity to ensure that the whale populations of the world are not endangered and to demonstrate that it wishes to conserve, control and manage the world's whale stocks in a wise and responsible manner.

Yes, all is far from well, but without the IWC it would be much worse. World attention is focussed on its debates and resolutions which are turning more and more to conservation, thanks to the new public awareness and sympathy for the plight of the cetaceans.

I have tried to make this book strictly factual and unsentimental, while yet my heart is torn with anxiety that these sentient, intelligent tribal groups of a unique and older civilisation than ours are being destroyed by the greed and cruelty of man. To me this destruction seems like genocide of our innocent, free and happy sea cousins.

My hope is that this book will enlist more support for saving the whales, the dolphins and the porpoises. I beg the reader to do all he or she can to help.

APPENDIX: Check list of living cetaceans of the world

The order *Cetacea* is divided into three sub-orders:

Mysteceti Baleen whales, from the Greek meaning 'moustached', referring to the hairy fringes of rod-like plates suspended from the upper jaws, and somewhat erroneously known in the trade as whalebone. The only true hairs are a few about the face and back. Two external nostrils or blow-holes. There are ten species. Strictly oceanic.

Odontoceti The toothed whales, dolphins and porpoises, at least 68 species occupying a wide diversity of habitat from the deep ocean to rivers hundreds of miles from the sea. Of these only the sperm or cachalot is as large as a baleen whale. The majority are less than 5m (16ft) in length. One blow-hole.

Archaeoceti Early cetaceans with elongated bodies and more primitive brains, known only from fossil material still being discovered.

Sub-order **Mysteceti**

Family BALAENIDAE

Balaena glacialis	Black right whale
mysticetus	Greenland right or bowhead whale
Caperea marginata	Pygmy right whale

Family BALAENOPTERIDAE

Balaenoptera acutorostrata	Minke whale
edeni	Bryde's whale
borealis	Sei whale
physalus	Fin whale
musculus	Blue whale
Megaptera novaeangliae	Humpback whale

Family ESCHRICHTIIDAE

Eschrichtius robustus	Gray whale

Sub-order **Odontoceti**

Family PLATANISTIDAE

Platanista gangetica	Gangetic dolphin
Inia geoffrensis	Bouto
Lipotes vexillifer	White flag river dolphin
Pontoporia blainvillei	La Plata dolphin

Family DELPHINIDAE

Steno bredanensis	Rough-toothed dolphin
Sousa teuszii	Cameroun river dolphin
plumbea	Plumbeous dolphin

lentiginosa	Speckled dolphin
borneensis	Borneo white dolphin
chinensis	Chinese white dolphin
Sotalia fluviatilis	Bufeo or tookashee
guianensis	Guiana dolphin
Tursiops truncatus	Bottle-nosed dolphin
Grampus griseus	Risso's dolphin
Lagenorhynchus albirostris	White-beaked dolphin
acutus	Atlantic white-sided dolphin
obliquidens	N. Pacific white-sided dolphin
australis	Black-chinned dolphin or Peale's porpoise
cruciger	Hourglass dolphin
obscurus	Dusky dolphin
Lagenodelphis hosei	Fraser's dolphin
Stenella longirostris	E. Pacific spinner dolphin
microps	Hawaiian spinner dolphin
attenuata	Spotted dolphin
caeruleoalba	Striped dolphin
Delphinus delphis	Common dolphin
Lissodelphis borealis	N. right whale dolphin
peronii	S. right whale dolphin
Cephalorhynchus commersoni	Commerson's dolphin
eutropia	White-bellied dolphin
havisidei	Haviside dolphin
hectori	Hector's dolphin
Peponocephalus electra	Broad-beaked dolphin
Feresa attenuata	Pygmy killer whale
Pseudorca crassidens	False killer whale
Globicephala melaena	Common pilot whale
macrorhyncha	Short-fin pilot whale
Orcinus orca	Orca or killer whale
Orcaella brevirostris	Irrawaddy dolphin
Phocaena phocaena	Common porpoise
sinus	California common porpoise
dioptrica	Spectacled porpoise
spinipinnis	Black porpoise
Neophocaena phocaenoides	Black finless porpoise
Phocaenoides dallii	Dall's porpoise
truei	True's porpoise
Family MONODONTIDAE	
Delphinapterus leucas	Beluga or white whale
Monodon monoceros	Narwhal
Family PHYSETERIDAE	
Physeter catodon	Sperm whale or cachalot
Kogia breviceps	Pygmy sperm whale

Family ZIPHIIDAE

Tasmacetus shepherdi	Tasman beaked whale
Mesoplodon bidens	Sowerby's whale
europaeus	Gulf Stream or Antillean beaked whale
mirus	True's beaked whale
pacificus	Longman's beaked whale
grayi	Scamperdown whale
hectori	Hector's beaked whale
stejnegeri	Bering Sea beaked whale or sabre-toothed whale
carlhubbsi	Arch-beaked whale
bowdoini	Andrew's whale
ginkgodens	Ginkgo-toothed whale
layardii	Strap-toothed whale
densirostris	Dense-beaked whale
Berardius arnouxi	S. giant bottle-nosed whale
bairdi	N. giant bottle-nosed whale
Hyperoodon ampullatus	N. Atlantic bottle-nosed whale
planifrons	Flat-headed bottle-nosed whale
Ziphius cavirostris	Cuvier's beaked whale

Acknowledgements and References

In preparing this book I have had the advice and help of many people in several capacities. Without giving a complete list I wish especially to acknowledge the assistance of Alan Best and David Sergeant of Canada; Charles and Virginia Jurasz of Alaska; Dr S. G. Brown and colleagues at the Whale Research Unit, and Dr Ray Gambell of the IWC (both now at Cambridge, UK); I. F. Lindsay, Dr L. Harrison Matthews, Frank Robson (New Zealand); Elizabeth Sutton; Tony Soper; and Drs F. C. Fraser and M. Sheldricks at the British Museum of Natural History.

The publications of The American Museum of Natural History and of the Canadian Sea Fisheries departments have been freely drawn upon for the results of present-day cetacean research, gratefully acknowleged, but far too numerous to list here. The following works have been a further source of information (published in London unless otherwise indicated):

Alpers, A. *A Book of Dolphins* (1960)
Bennett, F. D. *Narrative of a Whaling Voyage Round the Globe 1833–36* (1840)
Bruyns, W. J. F. M. *Field Guide of Whales & Dolphins* (Amsterdam, 1971)
Bullen, F. *The Cruise of the Cachalot* (1898)
Cousteau, J. Y. Recent books on whales and dolphins (1972–77)
Gambell, R. *Discovery* Reports (1968–72) and later papers to 1977
Gaskin, D. E. *Whales, Dolphins and Seals* (Auckland, 1972)
Hardy, A. *Great Waters* (1967)
Harrison, R. J. (editor) *Functional Anatomy of Marine Animals* (1972)
Lockyer, C. Journ Nat Hist 12: 513–528 (1978)
Matthews, L. H. (editor) *The Whale* (1968)
McIntyre, J. (editor) *Mind in the Waters* (New York, 1974)
Melville, H. *Moby Dick* (New York, 1851)
Mowat, F. *A Whale for the Killing* (New York, 1972)
Nayman, J. *Whales, Dolphins and Man* (1973)
Norris, K. S. (editor) *Whales, Dolphins and Porpoises* (Univ. of California)
Ommanney, F. D. *Lost Leviathan* (1971)
Ridgeway, S. H. (editor) *Mammals of the Sea* (Illinois, 1972)
Robson, F. *Thinking Dolphins, Talking Whales* (Wellington, NZ, 1976)
Schevill, W. E. (editor) *The Whale Problem* (Cambridge, Mass.)
Scoresby, W., jnr *An Account of the Arctic Regions, with a History and Description of the Northern Whale-Fishery* (Edinburgh, 1820)
Sergeant, D. E. Papers on sea-mammals of Canada (1975–78)
Slijper, E. J. *Whales* (New York, 1962)
Small, G. L. *The Blue Whale* (New York, 1971)
Stenuit, R. *The Dolphin, Cousin to Man* (1968)
Williams, H. (editor) *One Whaling Family* (1964)
Wood, F. G. *Marine Mammals and Man* (Washington, 1973)

Index

(For a full list of scientific names of cetaceans see pages 192-4; numbers in italic type indicate illustrations)

A porcelain model of a Fraser's dolphin Lagenodelphis Losei *with calf modelled by Elizabeth Sutton, the illustrator of this book*